INERA Conference: Vapor Phase Technologies for Metal Oxide and Carbon Nanostructures 2016

Journal of Physics: Conference Series Volume 764

Velingrad, Bulgaria
6 – 8 July 2016

Editors:

Diana Nesheva
Hassan Chamati
Julia Genova
Kostadinka Gesheva
Tatyana Ivanova
Albena Paskaleva
Anna Szekeres

ISBN: 978-1-5108-3276-3
ISSN: 1742-6588

Printed from e-media with permission by:

Curran Associates, Inc.
57 Morehouse Lane
Red Hook, NY 12571

Some format issues inherent in the e-media version may also appear in this print version.

Printed by Curran Associates, Inc. (2017)

Institute of Physics
Dirac House, Temple Back
Bristol BS1 6BE UK

Phone: 44 1 17 929 7481
Fax: 44 1 17 920 0979

techtracking@iop.org

Additional copies of this publication are available from:

Curran Associates, Inc.
57 Morehouse Lane
Red Hook, NY 12571 USA
Phone: 845-758-0400
Fax: 845-758-2633
Email: curran@proceedings.com
Web: www.proceedings.com

Table of contents

Volume 764

INERA Conference: Vapor Phase Technologies for Metal Oxide and Carbon Nanostructures

6–8 July 2016, Velingrad, Bulgaria

Accepted papers received: 4 October 2016
Published online: 15 November 2016

Preface

Papers

Lectures

Posters

PREFACE

We are pleased to introduce the Proceedings of the Conference "Vapor Phase Technologies for Metal Oxide and Carbon Nanostructures" (6th to 8th of July 2016, Velingrad, Bulgaria) organized by the Institute of Solid State Physics, Bulgarian Academy of Sciences in the frames of the Project INERA: "Research and Innovation Capacity Strengthening of ISSP-BAS in Multifunctional Nanostructures" REGPOT-2012-2013-1 NMP (http://www.inera.org). Participants from 7 different countries delivered 10 invited lectures, 16 oral and 26 poster presentations, contributing in 7 different topics. Papers submitted for publication in the Proceedings were refereed according to the standards of the Journal of Physics: Conference Series and the accepted ones illustrate the diversity and the high quality of the contributions. The Conference offered good opportunities for many interesting discussions and ample exchange of ideas between the participants. It became also a meeting point where INERA partners could plan their future collaboration and joint projects.

A significant factor for the success of the Conference was the conference venue – the beautiful spa resort Velingrad located 130 km from Sofia in the western part of the Rhodope Mountains. It is one of the leading "balneological" resorts in the Balkans with its ninety mineral water springs suitable for treatment of a wide range of diseases. The participants enjoyed also the rich social program, in particular the guided trip to the excavation site Dorkovo, one of the milestones marking the beginning of the Pliocene epoch in Eastern Europe.

The proceedings of conferences and workshops organized in the frames of INERA Project are regularly published by the *Journal of Physics: Conference Series*. We are grateful to the Journal's staff for the continuous support.

Managing Editor: Diana Nesheva

Co-editors: Hassan Chamati, Julia Genova, Kostadinka Gesheva, Tatyana Ivanova, Albena Paskaleva and Anna Szekeres

Journal of Physics: Conference Series **764** (2016) 011002 doi:10.1088/1742-6596/764/1/011002

INERA Conference 2016

Vapor Phase Technologies for Metal Oxide and Carbon Nanostructures

5-9 July, 2016, Velingrad, Bulgaria

http://inera.org

Chairman: Academician Alexander G. Petrov, INERA coordinator

Scientific Secretary: Professor Kostadinka Gesheva, INERA WP leader

Organizing committee:

Prof. Hassan Chamati – Director ISSP-BAS
Prof. Kiril Blagoev – INERA and ISSP-BAS Financial Director
Assoc. Prof. Emil Vlakhov – INERA Manager ISSP-BAS
Prof. Marin Gospodinov – ISSP-BAS
Assoc. Prof. Peter Rafailov – ISSP-BAS
Assoc. Prof. Yordan Marinov – ISSP-BAS
Assoc. Prof. Anna Szekeres – INERA ISSP-BAS
Assoc. Prof. Tatyana Ivanova – CLSENES-BAS
Assoc. Prof. Radostina Kamburova – ISSP-BAS
Dr. Dimitre Dimitrov – INERA ISSP-BAS
Dr. Jerome Leclercq – INERA ISSP-BAS
Dr. Blagoy Blagoev – INERA ISSP-BAS

Scientific committee:

Prof. Hassan Chamati – Director ISSP-BAS
Prof. Kiril Blagoev – NERA and ISSP-BAS Financial Director
Assoc. Prof. Emil Vlakhov – INERA Manager
Prof. Marin Gospodinov – ISSP-BAS
Prof. Nicholay Tonchev – ISSP-BAS
Assoc. Prof. Margarita Grozeva – ISSP-BAS
Assoc. Prof. Yordan Marinov – ISSP-BAS
Assoc. Prof. Emiliya Dimova – ISSP-BAS
Assoc. Prof. Victoria Vitkova – ISSP-BAS
Assoc. Prof. Albena Paskaleva – ISSP-BAS
Prof. Savcho Tinchev – IE-BAS
Prof. Petko Vitanov – CL SENES-BAS
Assoc. Prof. Georgi Popkirov – CL SENES-BAS
Assoc. Prof. Mariya Aleksandrova – TU-Sofia
Prof. Georgi Peev – UCTM
Prof. Frank Hamelmann – FT-BUAS (INERA partner), Germany

Journal of Physics: Conference Series **764** (2016) 011002 doi:10.1088/1742-6596/764/1/011002

Prof. Nicola Scaramuzza – DPC UC (INERA partner), Italy
Dr. Jan van Dijk – DAP TUE (INERA partner), Netherlands
Acad. Prof. DSc Henryk Szymczak – IP-PAS (INERA partner), Poland
Prof. Andrzej Wisniewski – IP-PAS (INERA partner), Poland
Prof. Andrzej Szemczyk – IP-PAS (INERA partner), Poland
Prof. Ion N. Mihailescu – NILPRP (INERA partner), Romania
Prof. Claes-Goran Granqvist – DES UU (INERA partner), Sweden
Prof. Mikael Jonsson – DES UU (INERA partner), Sweden
Prof. Andrew Livingston – DCECT IC (INERA partner), UK

Editorial committee:

Prof. Diana Nesheva (Managing Editor)
Prof. Hassan Chamati
Assoc. Prof. Julia Genova
Assoc. Prof. Tatyana Ivanova
Prof. Kostadinka Gesheva
Assoc. Prof. Anna Szekeres

Peer review statement

All papers published in this volume of *Journal of Physics: Conference Series* have been peer reviewed through processes administered by the proceedings Editors. Reviews were conducted by expert referees to the professional and scientific standards expected of a proceedings journal published by IOP Publishing.

Journal of Physics: Conference Series **764** (2016) 012001 doi:10.1088/1742-6596/764/1/012001

Thin film zinc oxide deposited by CVD and PVD

Frank U Hamelmann

Bielefeld University of Applied Sciences, Campus Minden, Artilleriestraße 9, 32427 Minden, Germany

E-mail: FHamelmann@fh-bielefeld.de

Abstract. Zinc oxide is known as a mineral since 1810, but it came to scientific interest after its optoelectronic properties found to be tuneable by p-type doping. Since the late 1980's the number of publications increased exponentially. All thin film deposition technologies, including sol-gel and spray pyrolysis, are able to produce ZnO films. However, for outstanding properties and specific doping, only chemical vapor deposition and physical vapor deposition have shown so far satisfying results in terms of high conductivity and high transparency. In this paper the different possibilities for doping will be discussed, some important applications of doped ZnO thin films will be presented. The deposition technologies used for industrial applications are shown in this paper. Especially sputtering of aluminium doped Zinc Oxide (ZnO:Al or AZO) and LPCVD of boron doped Zinc Oxide (ZnO:B or BZO) are used for the commercial production of transparent conductive oxide films on glass used for thin film photovoltaic cells. For this special application the typical process development for large area deposition is presented, with the important trade-off between optical properties (transparency and ability for light scattering) and electrical properties (conductivity). Also, the long term stability of doped ZnO films is important for applications, humidity in the ambient is often the reason for degradation of the films. The differences between the mentioned materials are presented.

1. Introduction

Zinc oxide is used as a material for centuries, known as a white powder and used as a pigment for paints ("chinese white"). Also for some other industrial applications zinc oxide is used. The electro-optical properties were first used for Ag based low emissive coatings in heat protective windows [1]. The common method for deposition of dielectric coatings for architectural coatings is reactive DC magnetron sputtering [2, 3].

Another important application of thin film zinc oxide films on large areas is using the fact, that ZnO is a wide band gap semiconducting material that can be doped. Good results can be found for n-type doping with aluminium, boron, gallium. Such doped ZnO can be used as transparent conductive oxide, especially important for thin film solar cells [4, 5, 6]. First results were obtained by Siemens Solar for the development of CIS (copper, indium, sulphur) based solar cells [7, 8]. Further development of this low pressure CVD was done at University of Neuchatel [4] with a commercial use by Oerlikon Solar, Switzerland [5]. In this case, the CVD precursors are diethyl zinc (DEZ) with water vapor for oxygen delivery. For doping, boron is introduced to form boron doped zinc oxide (BZO). The low deposition temperatures and high growth rates are suitable for production of TCO substrates with good transparency at low sheet resistance, and a unique film texture for efficient light trapping, which is needed for silicon based thin film solar cells.

The second technology for large area ZnO transparent conductive oxides is sputtering. The doping is usually done with aluminium (AZO). The development of high rate sputtering processes [9, 10] was commercialised by Applied Materials [11, 12]. For light trapping, the flat films were treated in a wet chemical etching process using HCl, craters are etched into the AZO film. Large area deposition up to six m^2 was achieved with this technology.

In this paper results of both technologies are shown: BZO and AZO films produced on industrial coating systems, and discuss the structural differences of the resulting films with a focus on the special features of BZO.

2. Method

The presented results have been measured in the laboratory on BZO and AZO films produced by production sized deposition systems on 1.4 square meter float glass substrates). The BZO films were deposited on the Oerlikon TCO-1200 LPCVD coating system, which a shortened version of the commercial product, consisting of a loadlock with pre-heating of the substrate, the deposition chamber with a three zone hot plate to keep the surface temperature constant on the whole substrate area, and an unloadlock for controlled cooling of the substrate. The temperature range used was between 150 and 190°C with a total pressure of typically 0,5 mbar. Mass flow controllers allow the flow of the precursors (DEZ and water vapor) through a showerhead. Diboran diluted in hydrogen was used for the doping. It must be noted that this system needs cleaning of the chambers, because some deposition occurs also on non-heated parts like the showerhead nozzles. To control the status of the chamber, a standard recipe was used before any other depositions. The commercial system consists of additional deposition chambers, so the throughput is multiple times higher.

The sputtered AZO films were deposited on a R&D sputtering system of Leybold Optics, suitable also for substrates up to 1.4 square meter. The system contains a loadlock for radiative preheating of the substrate, a deposition chamber for DC magnetron sputtering metal films (aluminium and silver) and a separate chamber for oxide films, including DC magnetron for AZO and RF magnetron for silicon oxide. An additional chamber allows several deposition runs at constant substrate speed (oscillating deposition). HCl etching can be perfomed in a modified glass washer constructed by RENA, containing a bath with 0,5% HCl solution. The speed of the substrate and the pH value of the solution and its temperature control the etching rate.

Films are characterized for thickness and sheet resistance on the whole substrate area with a self constructed x-y table measuring optical reflexion in the visible range and resistance with a 4-point probe. For further analysis, samples are cut into smaller pieces at several fixed spots of the full size substrate.

3. Results

A series of boron doped zinc oxide samples has been prepared with different depositon times, while all other deposition parameters were kept constant. The substrate temperature was fixed at 175°C, the DEZ/H2O ratio at 0.8 and the DEZ/B_2H_2 ratio was 0.25 (2% B_2H_2 diluted in hydrogen). The corresponding film thickness was 1.8 µm after 12 minutes, giving a deposition rate of about 2.5 nm/s. The linearity of the deposition after a short time for initial nucleation has been shown before [13]. The sheet resistance for the 1.8 µm film is about 7 Ω/sq. Figure 1 shows the development of the surface structure for each additional minute of deposition. The growth of the polycrystalline structure with pyramide shaped surface can be seen. From the SEM images, the average area of each pyramide was calculated, the square root gives the value of the surface feature.

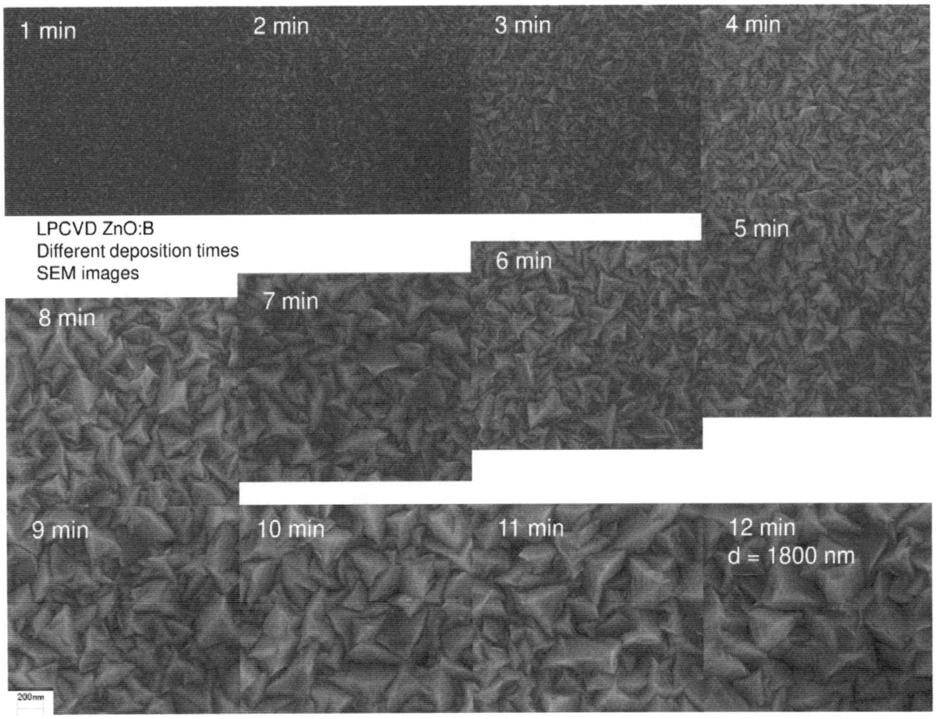

Figure 1. Development of BZO surface structure with time: SEM images.

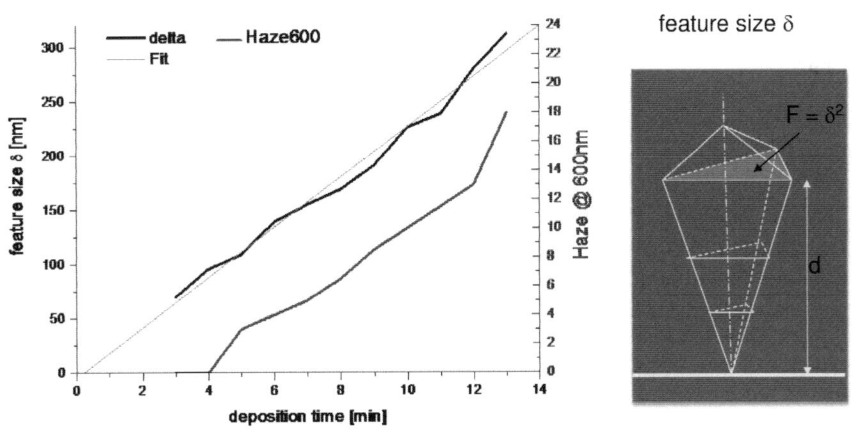

Figure 2. Correlation of surface feature size with haze.

Figure 2 shows the feature size as a function of deposition time, as it can be expected the correlation is linear. The graph of the haze at 600 nm wavelength as a function of film thickness is included as well, and shows a similar slope. The haze was calculated from transmission measurements of UV-vis spectrometer with and without integrating sphere. It gives the amount of diffuse scattered light to total transmitted light. This is an important value for the light scattering abilities of the film. However, it must be noted that the anglular distribution of the scattered light is important as well, but this haze does not take it into account.

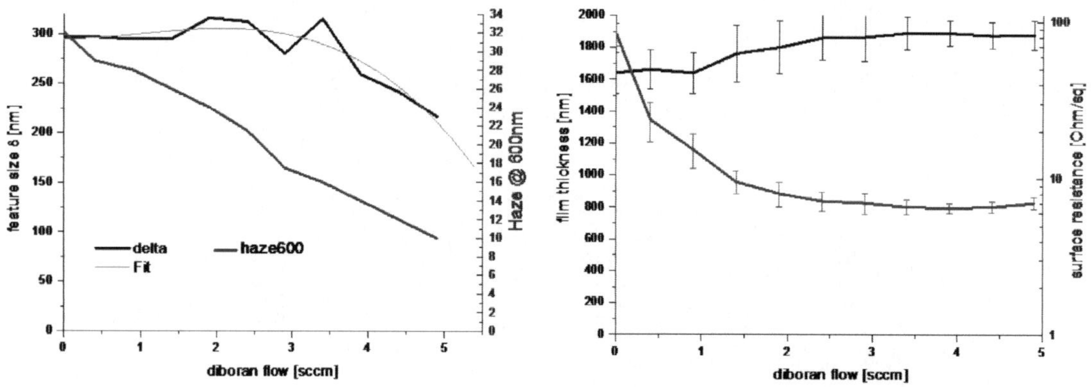

Figure 3. Feature size, haze, film thickness and sheet resistance as a function of diborane flow.

Another series of samples was prepared with variation of diborane flow in a wide range. As it can be seen in figure 3 (right), the sheet resistance is saturated at 3 sccm gas flow, while the film thickness is mainly uneffected. This can be explained by hall measurements that show the carrier concentration saturates, while the mobility of carriers decreases at higher levels of doping gas flow. This decrease in mobility may be explained by scattering at ionised impurities in the film [4]. However, from SEM images it can be seen that the feature size at high doping gas flows decreases slightly, while the haze decreases in a linear way from over 30% to 10% (left). To explain this, a rough estimation of the optical constants was done from reflexion R and transmission T values of the films. The results in figure 4 show in fact a strong influence of the optical constant from the doping gas flow. This has to be taken into account for all measurements based on the knowledge of the optical constants, like the determination of film thickness by optical reflectometry. For a more precise determination, spectral ellipsometric measurements would be useful.

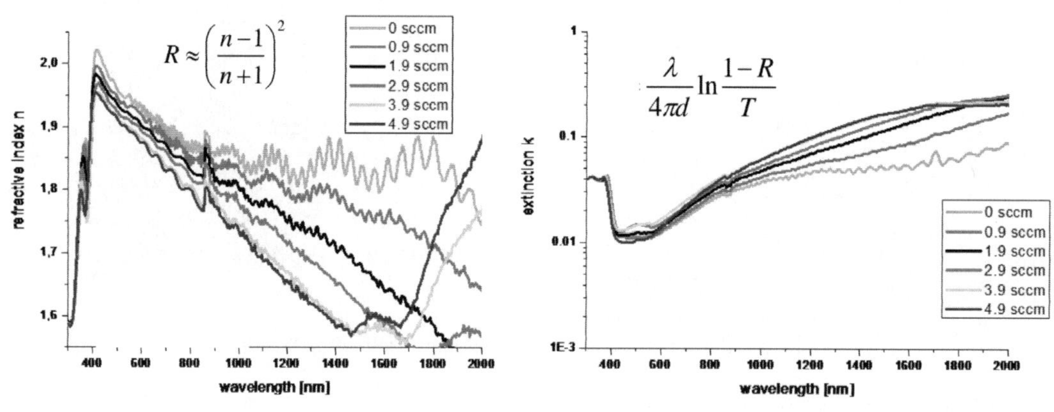

Figure 4. Calculated refractive index n and extinction k as a function of diborane flow.

While the BZO and AZO films are comparable in terms of transmission and sheet resistance [13], their structure is different. The orientation of the LPCVD produced BZO depends on the temperature, the conditions for the best growth of the material with respect to the application as TCO at temperatures around 175°C leads to c-axis parallel to the surface, (110)-plane is main direction as it can be seen in the XRD measurements shown in figure 5. The sputtered AZO shows clearly a different behaviour, the main direction is the (002)-plane, c-axis is perpendicular to the surface.

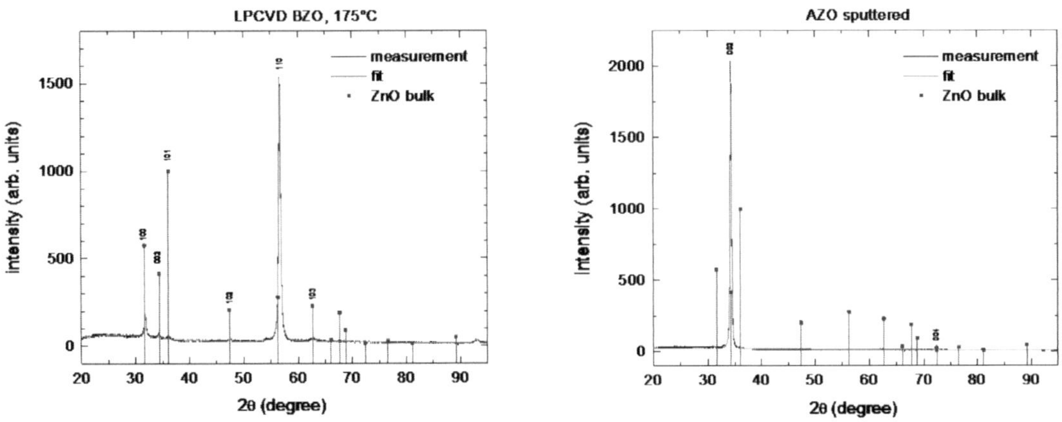

Figure 5. XRD results of BZO and AZO.

Another important difference is related to the light scattering abilities of the films. It must be noted, that the surface structure of the AZO films depends on the sputter parameters and the etching process. The results shown here were not optimized. As other authors have shown [14], this concept has room for further improvements. However, some of the developments in laboratory will be difficult to scale up for mass production. Figure 6 compares the angular distribution of the light scattering at 633 nm wavelength (left) and the haze (measured as mentioned before). To compare the results, three different commercially available TCO substrates with fluorin doped tin oxide are presented also. These commercial products are produced with an APCVD process in the float line of glass manufacturers. It can be seen, that the AZO has similar light scattering properties to the commercial SnO:F samples, the BZO has improved properties, the scattering is increased for all angles, while the AZO has only improved scattering in small angles, while it is reduced for the most important large angles. The resulting haze is higher for BZO in the whole for silicon solar cells relevant spectral area between 400 and 1100 nm.

Due to the structural differencies of BZO and AZO, the long term behaviour of the material is different. As published elsewhere [15], the BZO shows a connected network of voids, where diffusion of small ions can take place. It has also been shown [16], that exposure of the films to damp/heat (80% rel. humidity, 80°C for several hours) leads to an increased depletion layer that surrounds the crystals. Such behaviour could not be found for sputtered AZO. Such films proved to be stable in accelerated aging tests for a long time without any need for encapsulation [17]. Table 1 shows the results of a damp-heat test for 48 hours for different doped BZO films. It is found that highly doped films are more stable than low doped films.

Table 1. Sheet resistance of BZO before and after damp-heat test for 48 hours.

Sample	Resistance before DH	Resistance after DH
0 sccm Diboran	84 Ω/sq.	3.5 MΩ/sq.
0.9 sccm	15.3 Ω/sq.	1100 Ω/sq.
1.9 sccm	8.0 Ω/sq.	214 Ω/sq.
3.4 sccm	6.6 Ω/sq.	26 Ω/sq.
4.9 sccm	7.0 Ω/sq.	12.5 Ω/sq.

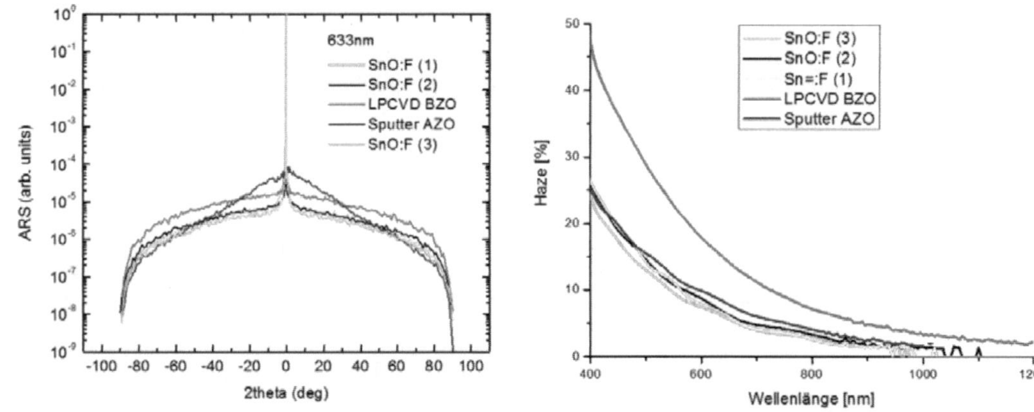

Figure 6. Angular distribution of light scattering (left) and haze at 600 nm wavelength (right).

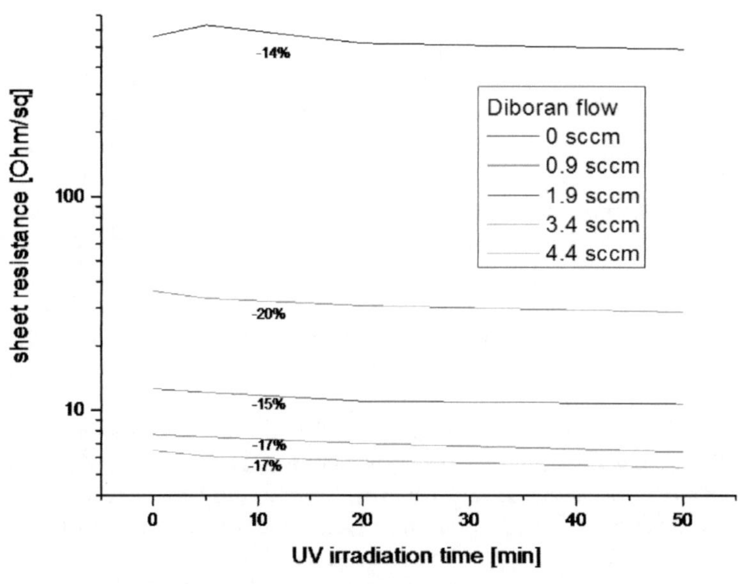

Figure 7. Variation of sheet resistance by treatment with UV light for different doped BZO films.

Another interesting effect is related to a post deposition treatment of BZO samples with a UV lamp with an emission of 185 and 254 nm. Such lamps are used for surface cleaning, since they also produce ozone in air ambient. A reasonable decrease of the sheet resistance could be found in samples with all doping levels. A lot of tests have been performed with different experimental conditions. It turned out, that the ozone production of the lamp does not influence this effect, since the UV treatment showed an even stronger decrease of the sheet resistance when the BZO sample was kept in vacuum. Also the addition of other gases to the process showed only a decrease of the effect, probably due to higher absorption of the UV light. The reduced sheet resistance turned out not to be completely stable, if the sample is heated to temperatures as used for solar cell deposition, the sheet resistance increases slightly but stabilized on a lower level compared to the as deposited values. It must be noted, that this effect does not occure on AZO samples, and also no effect could be seen for commercial SnO:F substrates. The mechanism of the UV treatment is still unclear, since the optical properties of the samples are not changing at all. Test of a UV treatment of BZO samples before a deposition of amorphous silicon solar cells showed a reasonable improvement of the fill factor, while open current voltage and short circuit current are not changed. The resulting relative improvement in efficiency was in the range of 10%, it is not clear if this is only a result of the lower sheet resistance, the contact between ZnO and silicon may also be improved by this treatment. The UV treatment has the potential to be included in a production process, it can be easily included in a deposition line at low costs. Further investigation can identify the process, and optimize the process with respect to the needed time.

4. Conclusions

Thin films of doped zinc oxide produced with deposition systems suitable for industrial mass production have been investigated. Boron doped zinc oxide (BZO) produced by LPCVD at 175°C shows some unique features for the use as transparent conductive oxide used for thin film silicon solar cells. The native surface texture that can be tailored by the deposition conditions results in excellent light scattering properties, in combination with high transparency and low surface resistance. The major disadvantage of this material is the sensitivity to humidity, for long term operation a good encapsulation is required, otherwise a degradation causing increased sheet resistance may be the result.

The structure of BZO is different from aluminium doped zinc oxide, deposited by DC magnetron sputtering. The AZO has a compact film structure, different crystal orientation. The as deposited films are flat. A surface roughness can be provided by an additional wet etching step in diluted HCl. This technology still has some potential for improvements, the surface structure can be tailored by sputter parameters and etching process details. The films are stable even in humid ambient.

An interesting effect was shown for BZO with post deposition UV light treatment. The sheet resistance of the film can be reasonable lowered by some irradiation of UV light. This effect has the potential to be included in mass production processes. Further investigation is required for this.

5. Acknowledgments

The author would like to thank his former colleagues from Malibu GmbH & Co. KG, especially H. Kurz, P. Prunici and T. Westerwalbesloh.

The author gratefully thanks the FP7 INERA project and its members from ISSP-BAS for the invitation and the perfect organisation.

References

[1] Gläser HJ 1980 *Glass Technol.* **21** 254
[2] Arbab M 2001 *Thin Solid Films* **381** 15
[3] Tsuda Y, Omoto H, Tanaka K, Ohsaki H 2006 *Thin Solid Films* **502** 223
[4] Fay S, Shaw A 2008 in: K. Ellmer, A. Klein, B. Rech, Zinc Oxide Grown by CVD as Transparent Contact for Thin Film Solar Cell Applications *Transparent Conductive Zinc Oxide, Springer Series in Material Science* **104**, 235 ed Ellmer K, Klein A, Rech B
[5] Kroll U, Meier J, Benagli S, Borello D, Steinhauser J, Vallat E, Castens L, Orhan J, Djeridane Y, Multone X, L. Fesquet L, Fecioru-Morariu M, B. Kalas J, Hoetzel J, P. Kupich M, O. Eisenhammer T 2011 *Proceedings 37th IEEE Photovoltaic Specialists Conference* 213
[6] Jost G, Merdzhanova T, Zimmermann T, Hüpkes T 2013 *Thin Solid Films* **532** 66
[7] Wieting RD, Potter RR. 1986 U.S. Patent 4,612,411
[8] Vijayakumar PS, Blaker KA, Wieting RD, Wong B, Halani A, Park C 1988 U.S. Patent 4,751,149
[9] Ellmer K, Kudella F, Mientus R, Schieck R, Fiechter S 1994 *Thin Solid Films* **247** 15
[10] Müller J, Schöpe G, Kluth O, Rech B, Sittinger V, Szyszka B, Geyer G, Lechner P, Schade H,

Ruske M, Dittmar G, Bochem HP 2003 *Thin Solid Films* **442** 158

[11] Klein S, Rohde M, Buschbaum S, Severin D 2012 *Solar Energy Materials and Solar Cells* **98** 363

[12] Klein S, Wieder S, Buschbaum S, Rohde M, Schwanitz K, Stolley T, Stömmer C, Severin D, Straub A, Schmidt UI, Ahmed K 2011 *Phys. Status Solidi* C **8** 2978

[13] Hamelmann F 2014 *Journal of Physics: Conference Series* IOP Publishing **559** 12016

[14] Dewan R, Shrestha S, Jordanov V, Hüpkes J, Bittkau K, Knipp D 2015 *Solar Energy Materials and Solar Cells* **143** 183

[15] Beyer W, Breuer U, Hamelmann F, Hüpkes J, Stärk A, Stiebig H, Zastrow U 2009 *MRS Proc.* **1165** M05-24

[16] Prunici P, Hamelmann F, Beyer W, Kurz H, Stiebig H 2013 *J. Appl. Phys.* **113** 123104

[17] Hüpkes J, Owen JI, Wimmer M, Ruske F, Greiner D, Klenk R, Zastrow U, Hotovy J 2014 *Thin Solid Films* **555** 48

Journal of Physics: Conference Series **764** (2016) 012002 doi:10.1088/1742-6596/764/1/012002

Thermochromic vanadium-dioxide-based thin films and nanoparticles: Survey of some buildings-related advances

Claes G Granqvist, Yu-Xia Ji, José Montero and Gunnar A Niklasson

Department of Engineering Sciences, The Ångström Laboratory, Uppsala University, P.O. Box 534, SE-75121 Uppsala, Sweden

E-mail: claes-goran.granqvist@angstrom.uu.se

Abstract. Today's architecture uses large glazings (windows and glass facades) to obtain good indoors–outdoors contact and day-lighting. However glazings offer challenges and often lead to excessive solar energy inflow and thereby a need for energy-demanding space cooling. This paper summarizes recent work on thermochromic (TC) materials intended for energy-efficient buildings and outlines how vanadium-dioxide-based thin films and nanoparticle composites can be used in TC glazings which admit more solar energy below a comfort temperature than above this temperature, so that the cooling need is diminished, while the transmittance of visible light remains high. We also report on some very recent work on TC light scattering.

1. Introduction

The carbon dioxide content in the earth's atmosphere keeps rising dramatically and has gone from ~315 ppm at the end of the 1950s to ~400 ppm in 2016; furthermore, the rate of increase has almost tripled during this period of time [1]. The growing amount of CO_2 is caused by unrestrained burning of coal, oil and gas, and is widely believed to be of great importance to life on Earth and cause global warming and rising sea levels [2]. The CO_2 level is amplified by increased plant respiration at elevated temperature [3,4]. There may also be many adverse secondary effects of climate change, *e.g.*, reduction of average global incomes [5] and climate-induced proneness to human conflict [6]. Aggravating factors are that the world's population continues to grow and is expected to be 50% larger in 2100 than today [7] and that the population is increasingly accumulated in mega-cities which act as urban heat islands with temperatures significantly above those of the surrounding countryside [8]. The human influence is so pervasive that the present geological era (Antropocene) is distinctly different from the era before the Industrial Revolution (Holocene) [9]. Given this background, it is evident that radical measures should be taken to decarbonize the energy sector, and this highlights the importance of buildings which currently are responsible for 30–40% of the world-wide use of primary energy [10].

Energy efficiency in the built environment is too often disregarded as an opportunity for CO_2 abatement [11] but, in fact, there are numerous emerging "green" technologies that can be harnessed [12-17], and energy efficient glazings (windows and glass facades) are particularly interesting. Glazings often allow too large energy flows to enter or exit a building so that energy-demanding cooling or heating become necessary, but small glazings are not a viable solution to the problem since precious indoors–outdoors contact and day-lighting are then sacrificed. However, energy efficiency can be reached with glazings that allow adjustable throughput of solar energy and visible light, and such glazings are often called "smart" or "intelligent". These glazings make use of chromogenic materials [18,19], and thermochromic (TC) ones are considered in this paper. TC materials have temperature-dependent properties, and thin films of them can transmit more solar energy at low temperatures than at high temperatures so that energy-efficient glazings can be accomplished and let in solar energy primarily when there is a need for such. TC-based fenestration was proposed almost 30 years ago [20,21] but has not yet made it to the marketplace—for reasons that are explained below—but there have been a number of recent advances that now seem to make practical implications much more likely [22-26]. The present paper is an update of some recent survey papers by one of the authors [27,28].

At this point it is appropriate to introduce some essentials of visible light (luminous radiation), solar irradiance and thermal radiation, which is done in figure 1. Thermal radiation, in panel (a), is governed by blackbody curves—which are shown for four values of the temperature τ—multiplied by a materials-dependent emittance which is less than unity; this radiation occurs at $\lambda > 2$ μm for ambient temperatures, where λ denotes wavelength. Solar radiation falling onto the earth's atmospheric envelope can be approximated by blackbody-like radiation corresponding to the sun's surface temperature (5505 °C) and lies at $0.2 < \lambda < 3$ μm. At ground level and under typical clear weather, this radiation is typically the one illustrated in figure 1(b). Finally, visible light is given by the bell-shaped curve in figure 1(b) which covers the $0.4 < \lambda < 0.7$ μm range and is peaked at 0.55 μm. Quantitative data on luminous (lum) and solar (sol) transmittance, denoted T_{lum} and T_{sol}, respectively, are derived by

$$T_{\text{lum,sol}} = \int d\lambda\ \varphi_{\text{lum,sol}}(\lambda)\ T(\lambda)\ /\ \int d\lambda\ \varphi_{\text{lum,sol}}(\lambda)\ , \tag{1}$$

where $T(\lambda)$ is spectral transmittance, φ_{lum} is the eye's sensitivity [29] and φ_{sol} is the "air mass 1.5" solar irradiance spectrum (for the sun standing 37° above the horizon) [30]. Analogous formulas pertain to reflectance R.

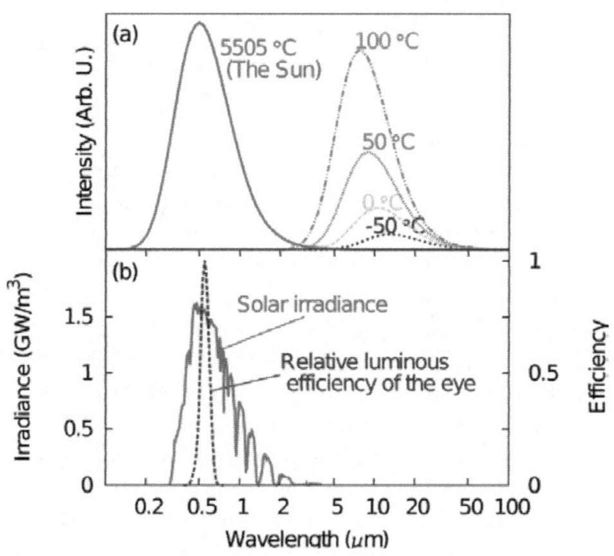

Figure 1. (a) Blackbody spectra for the shown temperatures, which include the surface temperature of the sun (the vertical scales differ for the various spectra). (b) Typical solar irradiance spectrum for clear weather (power per unit area and wavelength increment), and relative spectral sensitivity of the light-adapted human eye.

2. Vanadium dioxide thin films

It is convenient to introduce inorganic TC materials via their temperature-dependent electrical properties [20]. The conductivity changes reversibly and abruptly, sometimes by several orders of magnitude, at a well-defined "critical" temperature τ_c, the effect being due to a modification of the crystal structure. Vanadium dioxide (VO_2) is particularly interesting since τ_c lies at ~68 °C [31], which is not vastly different from ambient temperature meaning that this material may serve as a starting point for developing TC glazings of interest for buildings. VO_2 changes between a low-temperature ($\tau < \tau_c$) phase with semiconducting properties and high infrared ($\lambda > 0.7$ μm) transmittance as a thin film and a high-temperature ($\tau > \tau_c$) phase with metallic like properties and infrared reflectance. In fact, VO_2-based materials are the *only* inorganic materials considered to be likely candidates for implementation in TC glazings. The mechanism behind the reversible phase change at τ_c has been debated for decades [32] and is still not fully understood.

Figure 2 shows the most important optical properties of VO_2 thin films and nanoparticle deposits. The left-hand panels illustrate $T(\lambda)$ (upper) and $R(\lambda)$ (lower) for 50-nm-thick films in the spectral range of interest for solar radiation and demonstrate that $T(\lambda)$ is much larger in the semiconducting state at $\tau < \tau_c$ than in the metallic-like state at $\tau > \tau_c$ provided that $\lambda > 1$ μm; the difference in $T(\lambda)$ between low and high temperature is increased at longer wavelengths. Clearly VO_2 films have desired properties, in principle, and a TC glazing with such a film lets through more solar energy at a low temperature than at a high temperature [22-26,33-36]. Figure 2 also indicates that $R(\lambda)$ grows monotonically towards longer wavelengths for $\tau > \tau_c$, as expected for a metallic film. The right-hand panels in figure 2 will be discussed later.

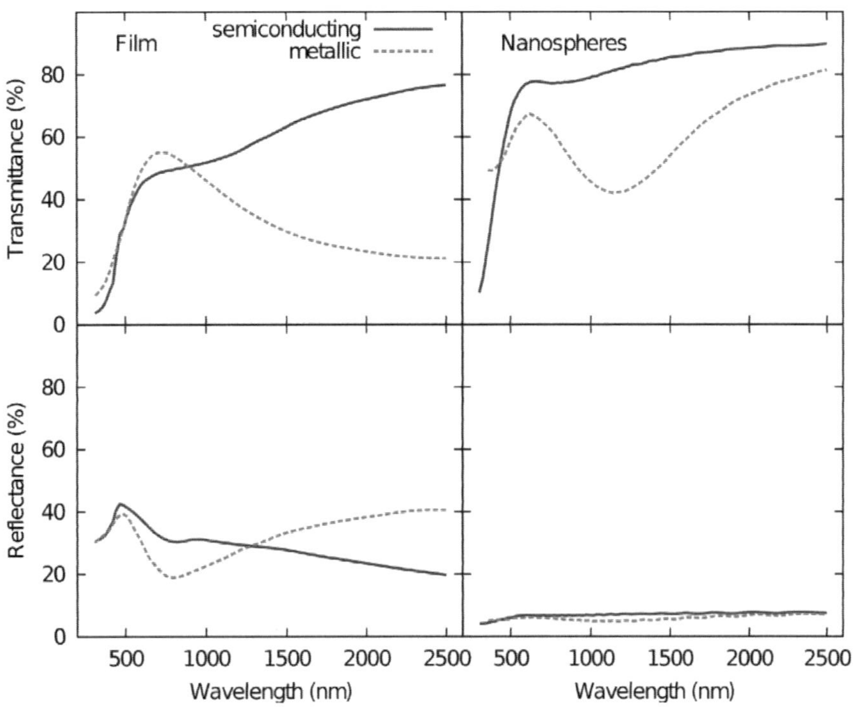

Figure 2. Spectral transmittance (upper panels) and reflectance (lower panels) for a 50-nm-thick VO_2 film (left-hand panels) and for a layer consisting of a dilute dispersion of VO_2 nanospheres, with an equivalent VO_2 thickness of 50 nm, in a medium mimicking transparent glass or polymer.

3. Towards practical VO_2-based thin films for glazings

The VO_2 films need to be modified in order to be of interest for practical glazings. Specifically, we consider three challenges [37]: (*i*) the change between semiconducting and metallic-like states occurs at ~68 °C which clearly is too high for buildings, (*ii*) according to figure 2, T_{lum} is as small as ~40%, which is too low for most practical glazings (a thinner film would yield a larger value of T_{lum} but the modulation of T_{sol} at τ_c would be very small), and (*iii*) transmittance modulation is pronounced only at wavelengths for which the solar irradiation is weak (see figure 1) so that the solar energy modulation is only ~10% and hence rather insignificant. These three challenges can be met, however.

Challenge (*i*) is easiest, and substituting some percent of the vanadium atoms by tungsten, to make $W_xV_{1-x}O_2$ films, allows τ_c to drop to ~25 °C or lower. Such films have been extensively studied [37], and data are consistent for well-crystallized materials. Several other additives can decrease τ_c, but W appears to be most efficient.

Challenge (*ii*) is connected with short-wavelength optical absorption, which is readily seen in figure 2. This absorption is caused by the undesirably narrow optical band gap in VO_2. Significant improvements can be achieved by adding a band-gap-widening element such as Mg [38,39], certain other alkaline earth metals [39], Zn [40] or Tb [41]. Fluorination, to make films of $VO_{2-z}F_z$, is another option for enhancing T_{lum} [42-44]. The effect of Mg addition is unambiguously caused by band gap widening, as evidenced by both experiments [45] and computations [46,47]. Additions of Mg or Tb lead to decreased values of τ_c, but the effect is not as strong as for W. It was found that T_{lum} could go from 39 to 51% when the Mg content was increased from zero to 7.2% [38].

The VO_2-based thin films must be prepared by a technology that permits large-area coatings, and several techniques involving physical (evaporation, sputter deposition) or chemical vapour deposition can be used, as reported in many publications. Sputter deposition may be of particular interest and is a standard technology for glass coating [48,49]. The substrate temperature typically should be ~450 °C [50] which is not a problem for thin films on glass whereas it offers difficulties for coating of plastic foil (which is an interesting technique since roll-to-roll coating can be particularly inexpensive [51]). Recent work on high-power impulse sputtering (HiPIMS) has demonstrated that the substrate temperature can be lower than for standard magnetron sputtering [52,53] but challenges remain. In general, the deposition of VO_2-based coatings must be judiciously controlled since the vanadium–oxygen phase diagram is exceedingly complex with the most oxygen-rich component being V_2O_5 [54,55]. Deposition in a reactive atmosphere of SO_2, rather than in a more common oxygen atmosphere, may be advantageous and permit a wider range of deposition parameters [26,56].

The fact that VO_2 is not the thermodynamically stable oxide is of interest also with regard to long term durability, which is an obvious requirement for thin films to be used in TC glazings. It has been observed that VO_2 films that are exposed to ambient air tend to gradually convert to V_2O_5, but over-coatings of Al_2O_3 or AlN can effectively prevent this process and impart long-term durability at elevated temperature and humidity [57,58].

4. Nanotermochromism in VO_2-based composites

The small modulation of the solar energy transmittance around τ_c - *i.e.*, item (*iii*) above - has been the most persistent problem for developing TC glazings. However it was recently discovered via calculations that a layer consisting of spheroidal nanoparticles of VO_2 - instead of a continuous VO_2 film—can improve the situation radically [59,60]. This kind of "nanothermochromism" was demonstrated in the right-hand panels in figure 2 above with regard to $T(\lambda)$ and $R(\lambda)$ for a 5-μm-thick layer of a material representative of typical glass or polymer with 1 vol.% of well distributed VO_2 nanospheres. This choice of parameters gives an equivalent VO_2 thickness of 50 nm so that it is possible to compare with the thin-film data also shown in figure 2.

It is now evident that several important features differ among thin films and nanoparticle composites of VO_2: (*i*) the nanoparticles composite is much more transparent than the corresponding thin film, (*ii*) metallic-like nanoparticles at $\tau > \tau_c$ display a deep transmittance minimum in the $0.7 < \lambda < 1.5$ μm range, and (*iii*) the nanoparticles absorb rather than reflect. Item (*ii*) implies that the nanoparticle composite is efficient for lowering the transmittance of solar energy whereas the luminous properties are not strongly influenced. The transmittance minimum corresponds to a peak in optical absorption and is caused by plasmon absorption [61], which occurs for a wavelength interval that is almost ideally located for controlling the solar energy inflow through a transparent TC glazing. The VO_2 nanoparticles must have sizes that are much smaller than the relevant wavelengths in order not to scatter light [62]. At larger sizes there can be pronounced TC light scattering, which is a newly observed phenomenon to which we return below.

Nanoparticle composites can be made by a variety of routes. Figure 3 reports some recent results on the nanostructure of a VO_2–SiO_2 composite prepared by reactive magnetron co-sputtering [63]. Scanning electron microscopy (SEM) displayed rounded nanoparticles with sizes of 100–300 nm and energy-dispersive x-ray (EDX) studies indicated that these particles were vanadium-rich. The films showed pronounced nanothermochromism in overall agreement with the results in figure 2(b).

(a) 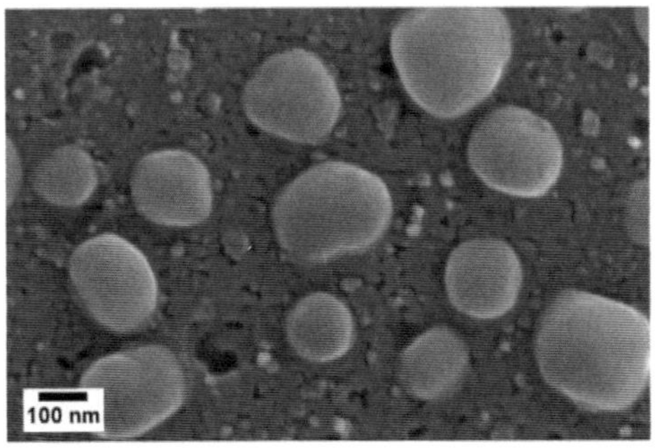

(b) 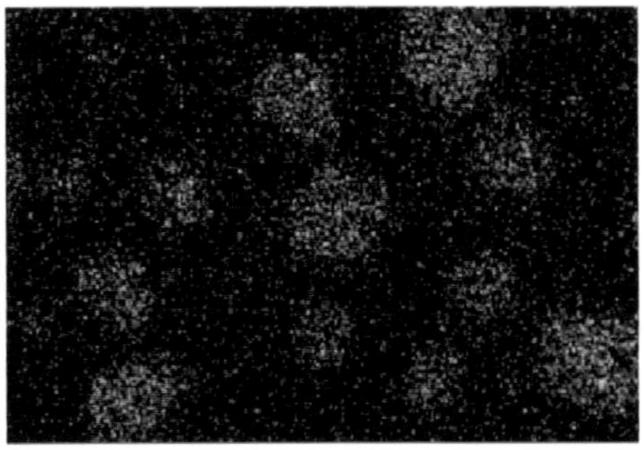

Figure 3. (a) SEM and (b) EDX data for a 220-nm-thick VO_2–SiO_2 film prepared by sputter deposition; green dots signify vanadium. From Reference 63.

5. Performance limits for thermochromic glazings

The work on VO_2-based thin films and nanoparticle composites that has been done during recent years makes it possible to state approximate performance limits for the luminous and solar properties of TC glazings [24]. These limits are reported in figure 4, specifically for T_{lum} at $\tau < \tau_c$ (see Equation 1) and for ΔT_{sol} defined by

$$\Delta T_{sol} = T_{sol}(\tau < \tau_c) - T_{sol}(\tau > \tau_c) \ . \qquad (2)$$

The values of T_{lum} are similar for $\tau < \tau_c$ and for $\tau > \tau_c$. It is desirable to have large magnitudes of both T_{lum} and ΔT_{sol}.

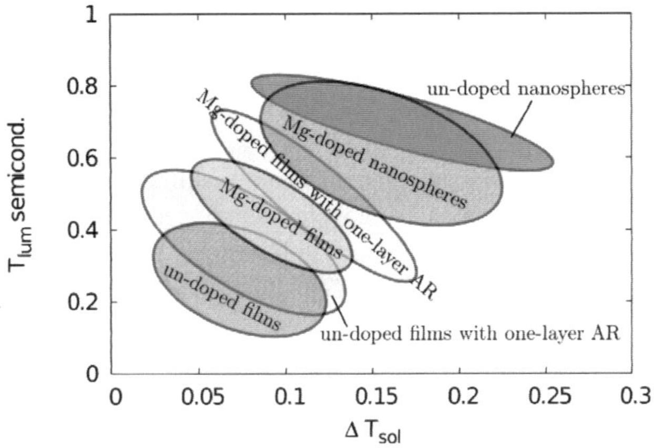

Figure 4. Approximate performance limits for thermochromic glazings incorporating VO_2-based materials. Data are shown for luminous transmittance at a low-enough temperature (semiconducting state) and for solar transmittance modulation. AR indicates antireflection. From Reference 24 (reproduced with permission from AIP Publishing LLC).

It is found that thin films of pure VO_2 have low values of T_{lum} and ΔT_{sol}, which highlights that these films are not of much interest for practical TC glazings. Antireflection layers can improve the performance, and particularly good data have been obtained with multilayers of TiO_2 and VO_2 [64,65]. Another possibility is to use "bio-inspired" cone-shaped surface structures to enhance the properties [66]. Mg-containing VO_2 thin films show even better properties, which can be boosted by antireflection treatment. However the best performance is obtained with VO_2 nanoparticles which are able to give $T_{lum} \approx 80\%$ together with $\Delta T_{sol} > 20\%$, and these properties are good enough to point at TC glazings as a viable option for practical implementation in buildings. Perhaps surprisingly, Mg-containing VO_2 nanoparticles do not give better results than pure VO_2 nanoparticles. The reason is that the Mg addition diminishes ΔT_{sol}, and this effect is larger than that of the band gap widening. Some small improvements can be achieved by optimized particle shapes and orientations [59] and by using core–shell structures [60].

6. Thermochromic light scattering

Temperature dependent light scattering is well known in Christiansen filters which combine materials with refractive indices that have strongly different temperature dependencies; these filters are clear when the refractive indices are matched whereas pronounced light scattering occurs when the indices are markedly different and the modulation of light scattering takes place gradually as the temperature is changed [67-69]. Light scattering from VO_2 particles has not attracted much attention in earlier work, but we recently performed an explorative investigation of a TC film with particles as large as ~1 μm [70]. This size is of the same order-of-magnitude as the wavelengths for visible light and solar radiation. The results of this study are summarized below. The applications of strongly scattering materials are not as straight-forward as for thin films and nanoparticle composites in TC glazings, but functional fibre mats [71,72] or other composites [73] may be areas where TC light scattering can be put to use.

Figure 5 shows the microstructure of a VO_2 layer made by sputter deposition onto a substrate that was pre-coated with In_2O_3:Sn (known as ITO) which promotes crystallinity and under certain conditions leads to the growth of particulate thin films [74]. The maximum peak height of the deposit was found to be ~190 nm. The structure seen in the atomic force microscopy (AFM) image was quantified with regard to particle volumes V_p and surface areas S_p using available computer code [75]. In order to enable a theoretical description of light scattering data, we modelled the individual particles by use of two different sets of equivalent sphere radii: either equal-volume spheres characterized by $r_{eq} = (3V_p/4\pi)^{1/3}$ or equal-volume-to-area spheres characterized by $r'_{eq} = 3V_p/S_p$. Prior work has shown that the latter measure can work well for representing light scattering from irregular objects [76]; the particle number is not conserved and there are more equivalent spheres than real particles. Figure 6 illustrates the corresponding radial distributions for r_{eq} and r'_{eq}.

Specular and diffuse transmittance were measured at $\tau < \tau_c$ and $\tau > \tau_c$ by spectrophotometry and there results were used to extract approximate absorption and scattering coefficients, denoted α_{abs} and α_{sca}, respectively; these data are shown in figure 7. Corresponding computed quantities, denoted K^*_{abs} and K^*_{sca}, respectively, were then derived by applying the Lorenz–Mie theory [77,78] to the particle size distributions in figure 6. It is now apparent from figure 7 that the calculations based on the equal-volume-to-area approximation can be reconciled with the experimental data at $\lambda < 0.7$ μm for absorption as well as scattering and at both low and high temperature. The equal-volume approximation clearly is less successful. Hence a semi-quantitative description of TC light scattering seems to be feasible even from highly irregular particles.

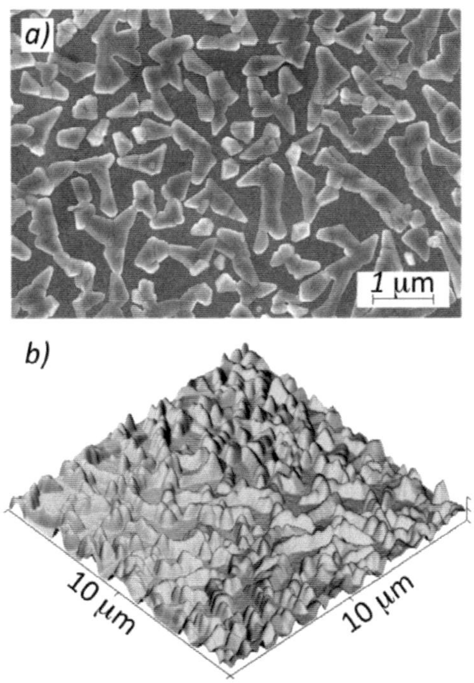

Figure 5. SEM (a) and AFM images (b) of a particulate VO_2 layer. From Reference 70 (reproduced with permission from AIP Publishing LLC).

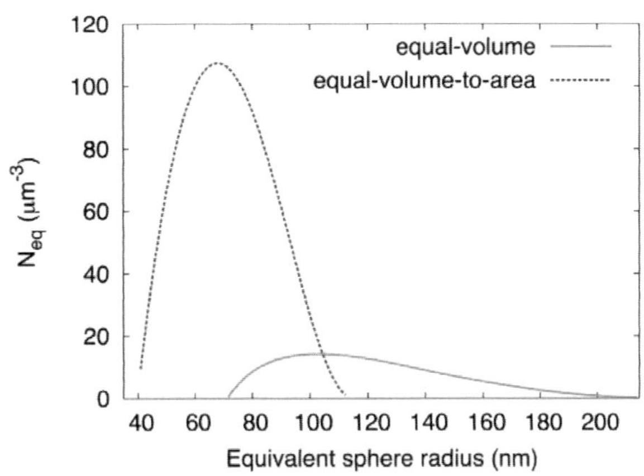

Figure 6. Distributions of the number N_{eq} of equivalent sphere radii for the data shown in figure 4(b). From Reference 70 (reproduced with permission from AIP Publishing LLC).

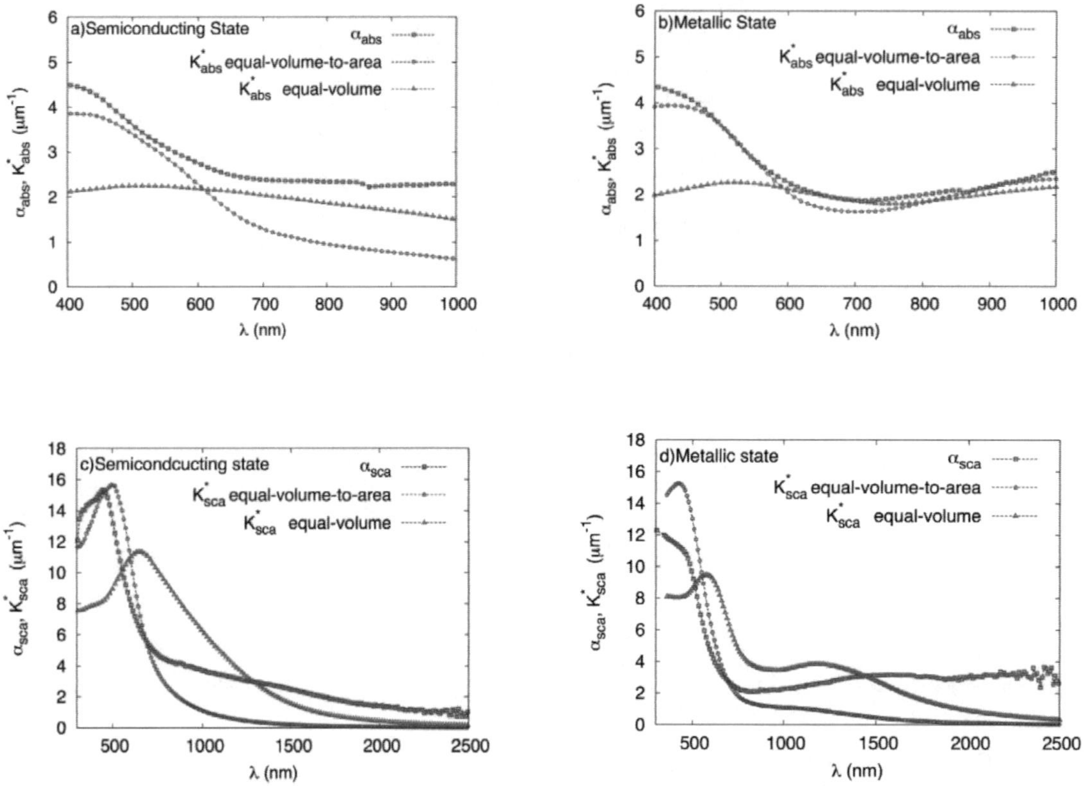

Figure 7. Experimental ($\alpha_{\rm abs}$) and theoretical ($K^*_{\rm abs}$) spectral absorption data, and corresponding results for experimental (α_{sca}) and theoretical (K^*_{sca}) spectral scattering data, for the VO_2 sample depicted in Figure 5. The calculations used the equal-volume and equal volume-to area approximations in figure 6. Data are given for the semiconducting state at $\tau < \tau_c$ [panels (a) and (c)] and the metallic state at $\tau > \tau_c$ [panels (b) and (d)]. From Reference 70 (reproduced with permission from AIP Publishing LLC).

7. Summary and conclusions

This article summarized a large body of recent computational and experimental results for thermochromic VO_2-based thin films and nanoparticle composites and discussed their potential for applications in glazings in energy-efficient buildings. It is our contention that such glazings can be made with properties that make them interesting for practical applications. In particular, nanothermochromism seems to offer new possibilities and VO_2 nanoparticles may be incorporated in functionalized glass laminates—perhaps even in the electrolyte layer in a combined electrochromic–thermochromic glazing. We also summarized recent work that indicates that thermochromic light scattering can be achieved and theoretically understood.

Acknowledgement

This paper was presented at the INERA Conference on "Vapor Phase Technologies for Metal Oxide and Carbon Nanostructures", July 5–9, 2016, in Velingrad, Bulgaria. The conference is part of the INERA REPGOT project of the Institute of Solid State Physics, Bulgarian Academy of Sciences. The work was financially supported by the European Research Council under the European Community's Seventh Framework Program (FP7/2007–2013)/ERC Grant Agreement No. 267234 ("GRINDOOR").

Journal of Physics: Conference Series **764** (2016) 012002 doi:10.1088/1742-6596/764/1/012002

References

[1] U.S. Department of Commerce, National Oceanic and Atmospheric Administration, Earthscan System Research Laboratory, Global Monitoring Division http://www.esrl.noaa.gov/gmd/ccgg/ trends/index.html
[2] Stocker T F, Qin D, Plattner G-K, Tignor M M B, Allen S K, Boschung J, Nauels A, Xia Y, Bex V and Midgley P M, editors (2013). *Climate Change 2013: The Physical Science Basis. Working Group I Contribution to the Fifth Assessment Report on the Intergovermental Panel on Climate Change* (Cambridge University Press, New York, NY, USA)
[3] Heimann M and Reichstein M 2008 *Nature* **451** 289
[4] Reich P B, Sendall K M, Stefanski A, Wei X, Rich R L and Montgomery R A 2016 *Nature* **531** 633
[5] Burke M, Hsiang S M and Miguel E 2015 *Nature* **527** 235
[6] Kelley C P, Mohtadi S, Cane M A, Seager R and Kushnir Y 2015, *Proc. US Natl. Acad. Sci.* **112**, 3241
[7] United Nations Department of Economic and Social Affairs 2015 *World Population Prospects: The 2015 Revision* (United Nations, New York, USA)
[8] Akbari H, Cartalis C, Kolokotsa D, Muscio A, Pisello A L, Rossi F, Santamouris M, Synnefa A, Wong N H and Zinzi M 2016 *J Civil Engr Management* **22** DOI: 10.3846/13923730.2015.1111934
[9] Waters C N *et al.* 2016 *Science* **351** 137
[10] United Nations Environmental Programme 2009 *Buildings and Climate Change: Summary for Decision-Makers* (UNDP Sustainable Buildings & Climate Initiative, Paris, France)
[11] Richter B, Goldston D, Crabtree G, Glicksman L, Goldstein D, Greene D, Kammen D, Levine M, Lubell M, Savitz M, Sperling D, Schlachter F, Scofield J and Dawson J. 2008 *Rev. Mod. Phys.* **80** S1
[12] Smith G B and Granqvist C G 2010 *Green Nanotechnology: Solutions for Sustainability and Energy in the Built Environment* (CRC Press, Boca Raton, FL, USA)
[13] Ginley D S and Cahen D, editors 2012 *Fundamentals of Materials for Energy and Environmental Sustainability* (Cambridge University Press, Cambridge, UK)
[14] Pacheco-Torgal F, Diamanti M V, Nazari A and Granqvist C G, editors 2013 *Nanotechnology in Eco-Efficient Construction* (Woodhead, Cambridge, UK)
[15] Pacheco-Torgal F, Mistretta M, Kauklauskas A, Granqvist C G and Cabeza L F, editors 2013 *Nearly Zero Energy Building Refurbishment* (Springer, London, UK)
[16] Pacheco-Torgal F, Labrincha J A, Cabeza L F and Granqvist C G, editors 2015 *Eco-Efficient Materials for Mitigating Building Cooling Needs: Design, Properties and Applications* (Woodhead, Cambridge, UK)
[17] Pacheco-Torgal F, Buratti C, Kalaiselvam S, Granqvist C G and Ivanov V editors 2016 *Nano and Biotech Based Materials for Energy Building Efficiency* (Springer International Publishers, Cham, Switzerland)
[18] Granqvist C G 1990 *Crit Rev Solid State Phys Mater Sci* **16** 291
[19] Lampert C M and Granqvist C G, editors 1990 *Large-Area Chromogenics: Materials and Devices for Transmittance Control*, SPIE Institutes for Advanced Optical Technologies, Vol. 4 (SPIE Opt Engr Press, Bellingham, WA, USA)
[20] Jorgenson G V and Lee J C 1986 *Sol Energy Mater* **14** 205
[21] Babulanam S M, Eriksson T S, Niklasson G A and Granqvist C G 1987 *Sol Energy Mater* **16** 347
[22] Gao Y, Luo H, Zhang Z, Kang L, Chen Z, Du J, Kanehira M and Cao C 2012 *Nano Energy* **1** 221
[23] Hoffmann S, Lee E S and Clavero C 2014 *Sol Energy Mater Sol Cells* **123** 66
[24] Li S-Y, Niklasson G A and Granqvist C G 2014 *J Appl Phys* **115**, 053513
[25] Warwick M E A and Binions R 2014 2014 *J Mater Chem A* **2** 3275

[26] Ji Y-X, Boman M, Niklasson G A and Granqvist C G 2016 in Pacheco-Torgal F, Buratti C, Kalaiselvam S, Granqvist C G and Ivanov V editors 2016 *Nano and Biotech Based Materials for Energy Building Efficiency* (Springer International Publishers, Cham, Switzerland) p 71
[27] Granqvist C G 2014 *J Vac Sci Technol* B **32** 060801
[28] Granqvist C G 2016 *Thin Solid Films* http://dx.doi.org/10.1016/j.tsf.2016.02.029
[29] Wyszecki J and Stiles W S 2000 *Color Science: Concepts and Methods, Quantitative Data and Formulae*, 2nd ed (Wiley, New York, NY, USA)
[30] ASTM 2008 G173-03 standard tables of reference solar spectral irradiances: Direct normal and hemispherical on a 37° tilted surface, in *Annual Book of ASTMStandards* (American Society for Testing and Materials, Philadelphia, PA, USA) Vol 14.04 http://rredc.nrel.gov/solar/spectra/am1.5
[31] Morin F J 1959 *Phys Rev Lett* **3** 34
[32] Weber C, O'Regan D D, Hine N D M, Payne M C, Kotliar G and Littlewook P B 2012 *Phys Rev Lett* **108** 256402
[33] Saeli M, Piccirillo C, Parkin I P, Binions R and Ridley I 2010 *Energy Buildings* **42** 166
[34] Saeli M, Piccirillo C, Parkin I P, Ridley I and Binions R 2010 *Sol Energy Mater Sol Cells* **94** 141
[35] Warwick M E A, Ridley I and Binions R 2014 *Energy Buildings* **77** 80
[36] Warwick M E A, Ridley I and Binions R 2015 *Sol Energy Mater Sol Cells* **140** 253
[37] Li S-Y, Niklasson G A and Granqvist C G 2012 *Thin Solid Films* **520** 3823
[38] Mlyuka N R, Niklasson G A and Granqvist C G 2009 *Appl Phys Lett* **94** 171909
[39] Dietrich M K, Kramm B G, Becker M, Meyer B K, Polity A and Klar P J 2015 *J Appl Phys* **117** 185301
[40] Jiang M, Bao S, Cao X, Li Y, Li S, Zhou H, Luo H and Jin P 2014 *Ceram Int* **40** 6331
[41] Wang N, Duchamp M, Dunin-Borokowski R E, Liu S, Zeng X T, Cao X and Long Y 2016 *Langmuir* **32** 759
[42] Khan K A and Granqvist C G 1989 *Appl Phys Lett* **55** 4
[43] Burkhardt W, Christmann T, Franke S, Kriegseis W, Meister D, Meyer B K, Niessner W, Schalch D and Scharmann A 2002 *Thin Solid Films* **402** 226
[44] Kiri P, Warwick M E A and Binions B 2011 *Thin Solid Films* **520** 1363
[45] Li S-Y, Mlyuka N R, Primetzhofer D, Hallén A, Possnert G, Niklasson G A and Granqvist C G 2013 *Appl Phys Lett* **103** 161907
[46] Hu S, Li S-Y, Ahuya R, Granqvist C G, Hermansson K, Niklasson G A and Scheicher R H 2012 *Appl. Phys Lett* **101** 201902
[47] Abdellaoui I, Merad G, Maaza M and Si Abdelkader H 2016 *J Alloys Cpds* **658** 569
[48] Pulker H K 1999 *Coatings on Glass*, 2nd ed (Elsevier, Amsterdam, The Netherlands)
[49] Gläser H J 2000 *Large Area Glass Coating* (von Ardenne Anlagentechnik GmbH, Dresden, Germany)
[50] Li S-Y, Namura K, Suzuki M, Niklasson G A and Granqvist C G 2013 *J Appl Phys* **114** 033516
[51] Bishop C A 2015 *Vacuum Deposition onto Webs Films and Foils* 3rd ed (William Andrew, Oxford, UK)
[52] Fortier J-P, Baloukas B, Zabeida O, Klemberg-Sapieha J E, Martinu L 2014 *Thin Solid Films* **125** 291
[53] Aijaz A, Ji Y-X, Montero J, Niklasson G A, Granqvist C G and Kubart T 2016 *Sol Energy Mater Sol Cells* **149** 137
[54] Wriedt H A 1989 *Bull Alloy Phase Diagr* **10** 271
[55] Kang Y-B 2012 *J Eur Ceram Soc* **32** 3187
[56] Ji Y-X, Niklasson G A, Granqvist C G and Boman M 2016 *Sol Energy Mater Sol Cells* **144** 713
[57] Ji Y-X, Li S-Y, Niklasson G A and Granqvist C G 2014 *Thin Solid Films* **562** 568
[58] Ji Y-X, Niklasson G A and Granqvist C G 2014 *J Phys Conf Ser* **559** 012005
[59] Li S-Y, Niklasson G A and Granqvist C G 2010 *J Appl Phys* **108** 063525

[60] Li S-Y, Niklasson G A and Granqvist C G 2011 *J Appl Phys* **109** 113515
[61] Lopez R, Haynes T E, Boatner L A, Feldman L C and Haglund Jr R F 2002 *Opt Lett* **27** 1327
[62] Laaksonen K, Li S-Y, Puisto S R, Rostedt N K J, Ala-Nissila T, Granqvist C G, Nieminen R M and Niklasson G A 2014 *Sol Energy Mater Sol Cells* **130** 132
[63] Ji Y-X, Niklasson G A and Granqvist C G 2016 *To be published*
[64] Mlyuka N R, Niklasson G A and Granqvist C G 2009 *Phys Status Solidi A* **206** 2155
[65] Chen Z, Gao Y, Kang L, Du J, Zhang Z, Luo H, Miao H and Tan G 2011 *Sol Energy Mater Sol Cells* **95** 2677
[66] Taylor A, Parkin I, Noor N, Tummeltshammer C, Brown M S and Papakonstantinou I 2013 *Opt Expr* **21** A750
[67] Christiansen C 1884 *Ann Phys Chem* **23** 298
[68] Clarke R H 1968 *Appl Opt* **7** 861
[69] Andersson A M, Niklasson G A and Granqvist C G 1987 *Appl Opt* **26** 2164
[70] Montero J, Ji Y-X, Granqvist C G and Niklasson G A 2016 *J Appl Phys* **119** 085302
[71] Li S, Li Y, Qian K, Ji S, Luo H, Gao Y and Jin P 2014 ACS *Appl Mater Interfaces* **6** 9
[72] Qian K, Li S, Ji S, Li W, Li Y, Chen R and Jin P 2014 *Ceram Int* **40** 14517
[73] Moot T, Palin C, Mitran S, Cahoon J F and Lopez R 2016 *Adv Opt Mater* to be published
[74] Montero J, Ji Y-X, Li S-Y, Niklasson G A and Granqvist C G 2015 *J Vac Sci Technol* B **33** 031805
[75] Nečas D and Klapetek P 2012 *Centr Eur J Phys* **10** 181
[76] Grenfell T C and Warren S G 1999 *J Geophys Res Atm* **104** 31697
[77] Mie G 1908 *Ann Phys* **330** 377
[78] Bohren C F and Huffman D R 1983 *Absorption and Scattering of Light by Small Particles* (Wiley, New York, USA)

Journal of Physics: Conference Series **764** (2016) 012003 doi:10.1088/1742-6596/764/1/012003

Some metal oxides and their applications for creation of Microsystems (MEMS) and Energy Harvesting Devices (EHD)

K Denishev

Technical University of Sofia, Faculty of Electronic Engineering and Technologies, Department of Microelectronics, 8, Kliment Ohridski Blvd, 1000-Sofia, Bulgaria

E-mail: khd@tu-sofia.bg

Abstract. This is a review of a part of the work of the Technological Design Group at Technical University of Sofia, Faculty of Electronic Engineering and Technologies, Department of Microelectronics. It is dealing with piezoelectric polymer materials and their application in different microsystems (MEMS) and Energy Harvesting Devices (EHD), some organic materials and their applications in organic (OLED) displays, some transparent conductive materials etc. The metal oxides Lead Zirconium Titanate (PZT) and Zinc Oxide (ZnO) are used as piezoelectric layers - driving part of different sensors, actuators and EHD. These materials are studied in term of their performance in dependence on the deposition conditions and parameters. They were deposited as thin films by using RF Sputtering System. As technological substrates, glass plates and Polyethylenetherephtalate (PET) foils were used. For characterization of the materials, a test structure, based on Surface Acoustic Waves (SAW), was designed and prepared. The layers were characterized by Fourier Transform Infrared spectroscopy (FTIR). The piezoelectric response was tested at variety of mechanical loads (tensile strain, stress) in static and dynamic (multiple bending) mode. The single-layered and double-layered structures were prepared for piezoelectric efficiency increase. A structure of piezoelectric energy transformer is proposed and investigated.

1. Introduction

Nowadays, the energy resources on the Earth are near to their depletion. For this reason, more attention has been started to pay to so called renewable energy sources and harvesting devices, to the decreasing of the power, consumed from various electronic devices, and to devices, having their own, autonomous power supply [1]. Many low power electronic devices, like clocks, portable radios, CD players, mobile telephones etc., have additional energy source, besides the battery supply [2]. The newly developed materials and the effects, related to them, play a key role to the design and the manufacturing of such type of devices.

Many microelectronic devices are used for generation of electricity, transducing different type of energy. It is related to the usage of specific materials, whose properties and effects provide electric charges, when mechanical stress or force is applied to them [3].

Piezoelectric effect is the ability of certain materials to generate an electric charge in response to applied mechanical stress. One of the unique characteristics of the piezoelectric effect is that it is reversible, meaning that materials, exhibiting the direct piezoelectric effect (generation of electricity, when stress is applied) exhibit the converse piezoelectric effect as well (generation of stress, when an electric field is applied) [4]. This effect is observed in crystals that have no center of symmetry. Some of the most popular materials, that possess piezoelectric properties, are lead zirconate titanate ($Pb(Zr,Ti)O_3$ or PZT), zinc oxide (ZnO), aluminum nitrate (AlN) etc. [5]. They show piezoelectric effect with different piezoelectric coefficients. The application of the piezomaterials is related to the knowledge of their parameters and the methods, used for their deposition [6].

Journal of Physics: Conference Series **764** (2016) 012003 doi:10.1088/1742-6596/764/1/012003

Some ceramic materials have dielectric constants, which are up to 100 times higher than these of the ordinary crystals. These synthetic materials are classified in a new group, that of ferroelectrics. They have piezoelectric constants much higher than that of the natural materials. This leads to intensive investigations for synthesizing of materials like PZT, which have favourable piezoelectric parameters, as well as that of zinc oxide (ZnO) [7].

Piezoelectric materials, such as PZT and ZnO are very important for the Micro-electro-mechanical systems (MEMS) and Nano-electro-mechanical systems (NEMS) design, as sensors and actuators. Traditionally, the PZT and ZnO are used as bulk materials, with thickness, higher than 100-200 μm. The requirements for miniaturization can be fulfilled by conversion of the piezoelectric materials into thin nanometric films, via using of some of the conventional microelectronic technologies, like RF sputtering.

2. Physical parameters of PZT and ZnO

Actually, **PZT** is piezoelectric ceramic, for which, the proportion between zirconium and titanium (Zr/Ti) is defined for certain specific application. There could be seen variety of compositions, but the most frequently used are 54/46; 52/48; 50/50; 53/47; 80/20 etc. PZT is a ferroelectric material which has excellent piezoelectric and pyroelectric properties. [8]. The synthesis of $Pb(Zr,Ti)O_3$ is rather complicated. The technological processes are carried out in tightly closed containers or capsules at high temperatures [9]. It should be mentioned that, the most of the available ceramics, such as PZT, are based on the structure of perovskite (figure 1). Usually, the typical thickness of the piezoelectric PZT layers in MEMS technologies is in the same range as that of ZnO layers - 0.05-3 μm [10].

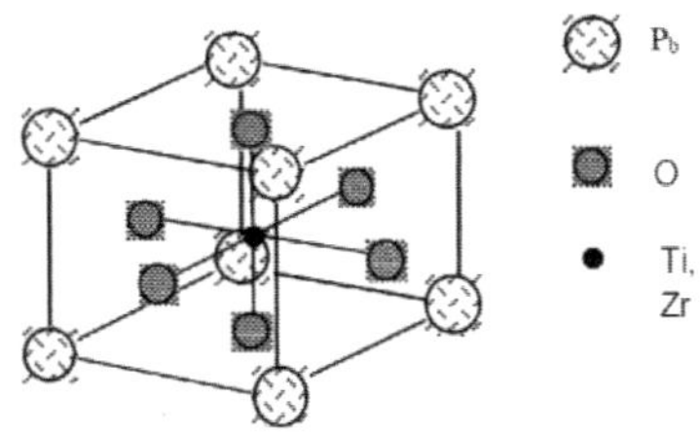

Figure 1. Perovskite structure of PZT.

The most important coefficients, related to piezoelectric films, and their values, typical for PZT nanometric films, are as follows:

- Piezoelectric coefficients (Strain)	d_{31}=180.10^{-12} C/N; d_{33}=360.10^{-12} C/N;
- Piezoelectric coefficients (Voltage)	g_{31}=0,011Vm/N; g_{33}=0,025 Vm/N;
- Coefficient of electromechanical transformation	k=0,35-0,69;
- Elasticity modulus	Y=49-63 GPa;
- Relative dielectric permittivity	ε_r=1700
- Currie temperature	T_C=330-360^{0}C [11].

ZnO or zinc oxide forms a hexagonal wurtzite structure with lattice constants a = 3.25 Å and c = 5.2 Å. It possesses the lack of center of symmetry, required for piezoelectric materials (figure 2). The piezoelectric coefficients for ZnO are also relatively high, which makes it an excellent material to be used in a wide variety of piezoelectric applications [12]. Besides in MEMS, they are used also in optoelectronics, in mobile communications for different kinds of filters, using surface acoustic waves and for thin film resonators for bulk acoustic waves. They are used also as chemical sensors.

The coefficient of electromechanical transformation of ZnO has relatively high value, which makes it an excellent material for large number of piezoelectric systems (MEMS, sensors and actuators). The properties of zinc oxide coatings depend on the method and parameters of deposition. RF Sputtering is the most frequently used deposition method due to the uniform thickness of the produced layer and because of the fact that it could be applied on different type of substrates.

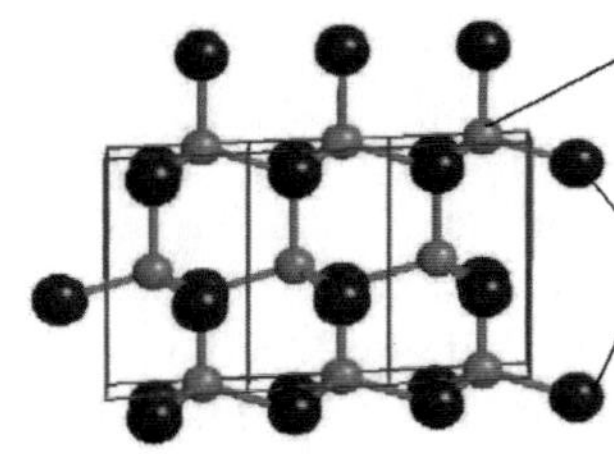

Figure 2. Hexagonal crystal structure of ZnO.

The typical values of the piezoelectric coefficients for ZnO nanometric films are as follows:

- Piezoelectric coefficients (Strain)	d_{31}=5.10^{-12} C/N; d_{33}=12,4.10^{-12} C/N;
- Piezoelectric coefficients (Voltage)	g_{31}=0,36 Vm/N; g_{33}=1,57 Vm/N;
- Coefficient of electromechanical transformation	k=0,33 [13];
- Elasticity modulus	Y=30-200 GPa;
- Relative dielectric permittivity	ε_r=10-11

3. Experimental work on PZT and ZnO layers deposition, their characterization and measurement of the SAW test structures

In our work these basic materials are used like thin films, with thickness of around x.100 nm. As a result the piezoelectric response is weaker, comparing to bulk material. Several groups of experiments were done with the aims, firstly, to estimate the quality and to determine the parameters of PZT and ZnO layers and on the second place, to define the proper deposition procedures for optimization of the layers' piezoelectric response. The films were deposited by using a RF Sputtering System. More details about the concrete deposition conditions can be found elsewhere [6, 8, 10]. For investigation of the films, a Surface Acoustic Wave (SAW) structure was prepared (figure 3), by using of which, the piezoelectric effects in the two materials, can be easily detected.

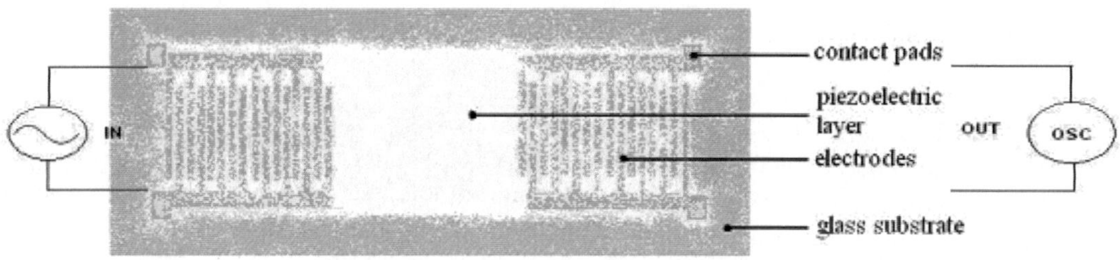

Figure 3. Top view of a SAW structure.

The structure consists of two pairs of comb electrodes, which are the input and output transducers, respectively. They were prepared by using of chrome layer, deposited on glass substrate and patterned by photolithography. The both samples (PZT and ZnO) were prepared with the same mask configuration and dimensions. The layers of piezoelectric material were deposited on top and between the electrodes. The aim is, by making of the input transducer to vibrate, a mechanical wave in the piezoelectric material to be generated. The vibration is excited by sinusoidal signal, provided by a functional generator (MPF3060, 60MHz) at certain frequencies. When a mechanical wave was generated, it was spread into the whole piezoelectric layer, deposited on the electrodes and the area between them. The wave induces a voltage between the electrodes, when it reaches the output transducer. By dual channel oscilloscope (DQ2041CN), the signals of the structure's input and output were observed simultaneously. In figure 4, the curves of ZnO and PZT reactions are shown [14].

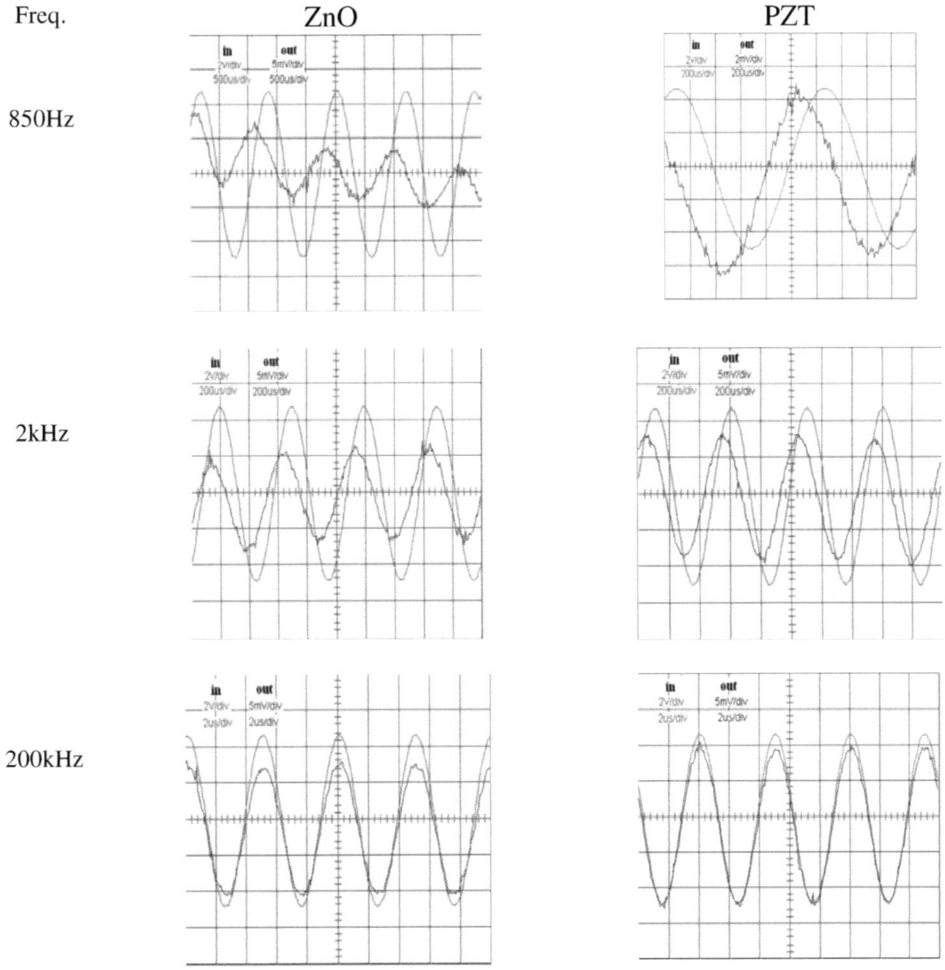

Figure 4. Voltage generated by the PZT and ZnO based SAW structures [14].

Because of the low elasticity of the glass substrate we cannot estimate the exact electromechanical coupling coefficients. This is the reason, the mechanical waves, induced by the input transducer to be restricted in their magnitude. As a consequence, the amplitude of the output voltage was also limited, but the results of 17 mV for ZnO and 10mV for PZT from peak to peak at input signal parameters 200kHz and 9Vp-p is quite good result for this film piezoelectric structure. The conclusion is that there is a coupling between the input and output transducers of the SAW structure, because of the presence of the piezoelectric materials. Moreover, the structure may serve as piezotransformer, as the output signal remains with sinusoidal shape with negligible distortions and no shift at the higher frequency (figure 4).

For better elasticity, a flexible organic substrate of polyethylenetherephtalate (PET) was used. In figure 5, a simplified technological sequence and the cross-section view of the structure is shown. For the top and bottom electrodes, Al layers, with thickness of 200 nm, were used. For the proposed structure, a piezoelectric material PZT, with a thickness of about ~150 nm and piezoelectric coefficient d_{33}=360 pC/N was used. By this approach, a bigger amplitude of the output voltage, in comparison to the ZnO (d_{33}=12.4 pC/N), was achieved.

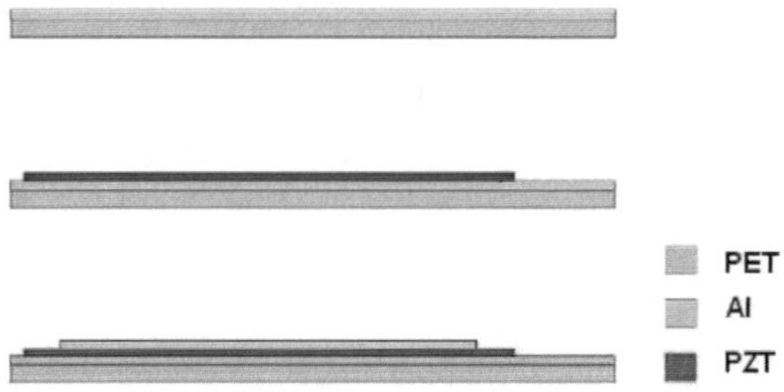

Figure 5. Technological sequence of microgenerator

For mechanical testing of the piezoelectric microgenerator in dynamic mode, the setup, presented in figure 6, was used. As a mechanical wave source an electromagnetic shaker was used. It consists of a coil with an anchor, at the end of which the sample is clipped. The coil was supplied by sinusoidal voltage with defined frequency. During the mechanical vibration of the structure, a voltage with almost the same sinusoidal shape was generated. The measurement was performed by using of a digital oscilloscope, whose two channels, were connected respectively to the sample and to the coil. Figure 7 represents the dependence of the generated voltage on the electrode's area, respectively for 100 Hz and 200 Hz. As can be seen, the increase of the electrodes area and the vibration frequency increases the output power of the generator, because of the fact that the flexible substrate provides higher elasticity of the whole structure, which ensures the corresponding voltage generation, by using of lower mechanical loading.

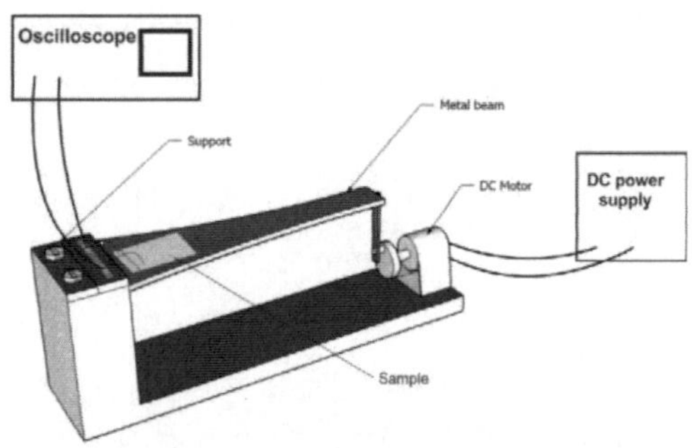

Figure 6. Measurement setup for dynamic mode testing.

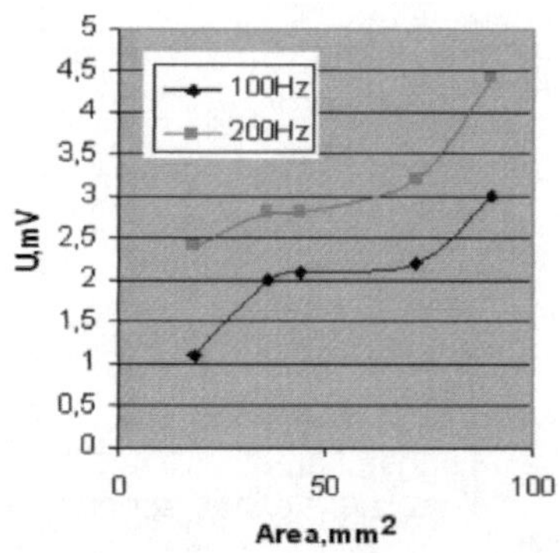

Figure 7. Dependence of the generated voltage versus the electrode's area at different activation frequencies.

In one of the studies of 180 nm PZT layers on flexible PET substrate, the effect of the aluminum electrodes configuration on the generated voltage at static stress was investigated. Both types of fabricated structures are shown in figure 8. Testing conditions and schematic view of the test setup for voltage-stress characteristic measurements can be found elsewhere [6]. The first one is sandwich type with top and bottom aluminum electrodes, deposited by thermal evaporation. The second configuration consists of PET substrate, layer PZT on it and finally – several top electrodes, laterally patterned.

In the case of the sandwich type structure (figure 8a), the sensor is not sensitive to mechanical stress, lower than 100 mg (figure 9). Possible reason for this lack of sensitivity, to weak stress, is the fact that the orientation and polarization of PZT particles, for this type of structure, are predominantly in lateral direction. As a result, the piezoelectric reaction is hard to be detected, with vertical loading of the electrodes. That's why a lateral structure was produced (figure 8b) and tested (figure 10).

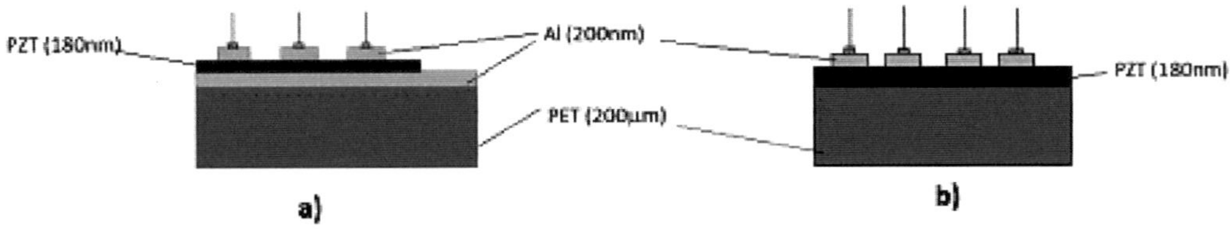

Figure 8. Cross section of the PZT piezoelectric sensor. a) sandwich type; b) lateral type [6].

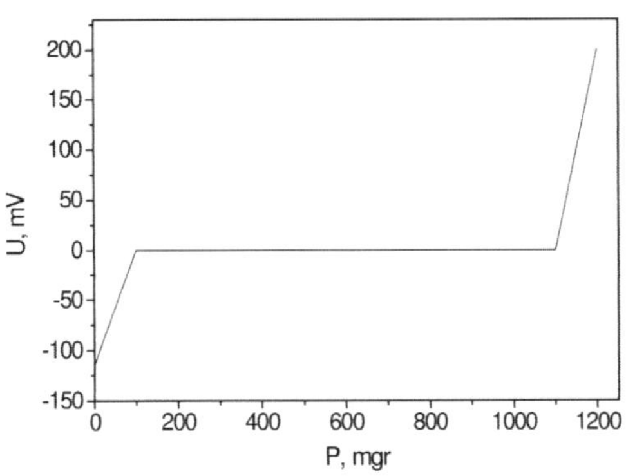

Figure 9. Voltage-stress dependence for the sandwich type structure [6].

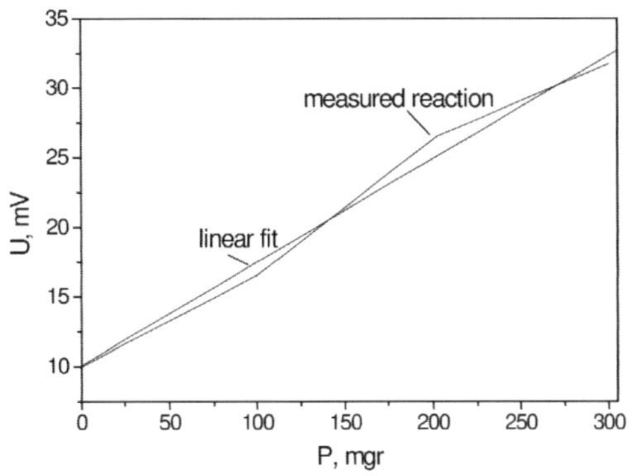

Figure 10. Voltage-stress dependence for the lateral type structure [6].

The piezoelectric coefficient, generating charge in direction, perpendicular to the applied stress, could be described by the equation [15]:

$$d_{31} = 2\varepsilon_0 \varepsilon_r P_r Q_{12}, \qquad (1)$$

where P_r is the voltage, measured during the experiment, ε_r is PZT dielectric permittivity (ε_r = 1700), ε_0 is dielectric constant in vacuum ($\varepsilon_0 = 8{,}85.10^{-12}$ F/m) and Q_{12} is the electrostrictive coefficient, or the ability of the dielectric materials to reorient the electrical domains inside the material. After calculations, for this coefficient, it was obtained value of d_{31}=186 pC/N. For comparison, the maximal value of this coefficient, for PZT-layers, deposited from solution, is 78 pC/N. This result can be explained by the better PZT layer morphology here. For the samples with planar electrodes, the sensitivity is higher than the values, obtained from the sandwich type structure and the voltage-stress characteristic is highly linear. The main reason is probably the fact that, during PZT deposition, the polarization is directed perpendicularly to the applied stress (d_{31}) [10]. In this case the charge is generated between two neighboring electrodes and can be easily measured. The sensor has relatively wide dynamic range, in which the linearity deviation is only 1.8 %.

The electrical parameters and reaction of PZT sensors with planar electrodes are quite similar and even higher than those, given in other publications. The eventual reasons for the good efficiency are:

1) the flexible substrate, providing higher mechanical deformation than silicon wafer, and 2) suitable deposition rate of sputtered PZT layers, allowing to the particles to take more favorable energetic positions and to build up proper crystal lattice with preliminary defined piezoelectric properties.

The next set of experiments was conducted to check and prove the microstructure and the state of the chemical bonds in the layers. For this reason, FTIR spectroscopy was applied with IRPresige 21 Shimadzu spectrophotometer in the spectral range of 350 cm^{-1} to 4000 cm^{-1}. FTIR spectroscopy as a very sensitive tool can give some information for the chemical bondings and structure of the studied samples.

The Infrared spectrum of **ZnO** films was obtained in reflectance mode in the spectral range of 350 to 1100 cm^{-1} (see figure 11). The absorption band at 390 cm^{-1} is usually reported for ZnO as a typical absorption feature [16]. The absorption peak at 406 cm^{-1} is theoretically confirmed for wurtzite zinc oxide as infrared active mode. The absorption band at 430 cm^{-1} is assigned to the stretching vibrations of Zn-O bond [17, 18], which is characteristic for ZnO. The IR lines at 560 and 814 cm^{-1} are also due to the stretching vibrations of Zn-O bond [19].

PZT have crystal structures belonging to the perovskite family with the general formula ABO_3. It is known that the perovskite phase is the only desired crystal structure which has the ideal piezoelectric properties. FTIR spectrum of PZT thin film, in reflectance mode, is presented in figure 12. The bands appearing in the range of 450 to 350 cm^{-1} are related to B–O vibrations (BO_6 and B–O for ABO_3 structures) [20, 21]. The distinctive feature at 550 cm^{-1} is observed. This band resembles the reported FTIR observations of PZT ceramics [22]. In this spectral area, a broad band can be seen from 800 to 550 cm^{-1} with a maximum at 780 cm^{-1}. This peak has been associated with the vibration of M–O (M = Zr or Ti) bonds in the systems.

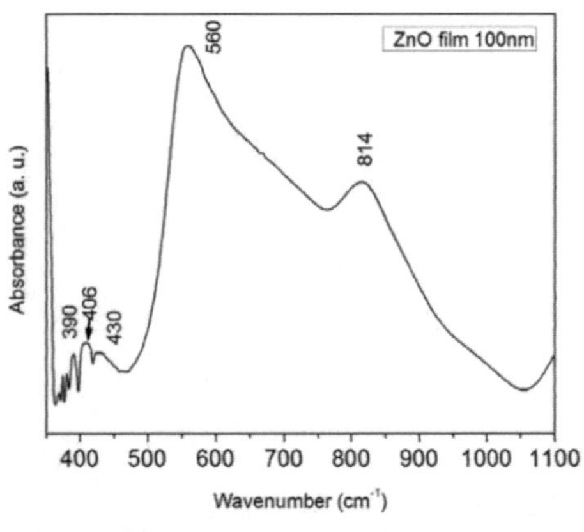

Figure 11. FTIR spectrum of sputtered ZnO film [23].

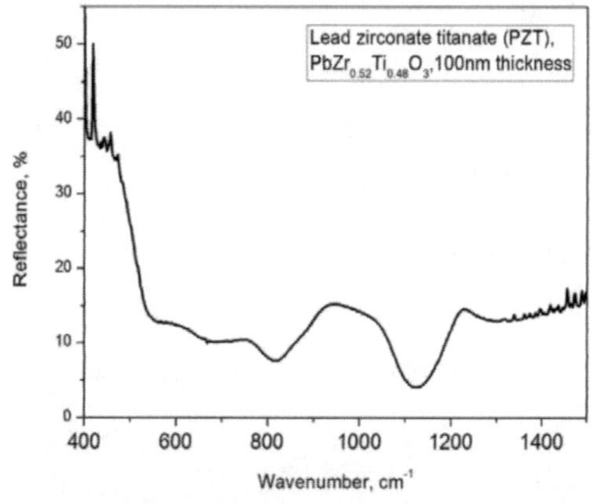

Figure 12. FTIR spectrum of sputtered PZT film [23].

The obtained FTIR results confirmed the formation of piezoelectric phase of ZnO and PZT. This is a promising result, as it manifests, that both films can be deposited by RF sputtering process and respectively these are favorable phase structures for exhibiting good piezoelectric properties.

4. Study of double-layer piezoelectric structures based on PZT and ZnO.

This section presents the study of double-layer flexible piezoelectric structures, containing combination of PZT and ZnO single-layers in different order of deposition in relation to the near substrate criteria. Firstly, PZT was deposited on the PET, covered by Al, covered as a second layer of ZnO, and vice versa. The aim of this study was fabrication of piezoelectric transformers. For comparison data about the single layered structures are given. Oscillograms of the signal produced are presented in figures 13 and 14 [24].

For the **single layer** configuration the voltage, generated from the 200 nm thick PZT layer was in the range 40-60 mV. The values of 40 mV were measured at 23 Hz and 33 Hz and the values of 60 mV at 45 Hz and 55 Hz. At the same measuring conditions, the generated voltage, taken from the single-layer piezoelectric structure of ZnO was in the range 80-100 mV - 80 mV was measured at 23 Hz, 45 Hz and 55 Hz and 100 mV, was measured at 33 Hz.

For the **double layered structure** with first layer of PZT and second ZnO, the generated piezoelectric voltage is almost a half of that, obtained from the previous one-layer structures. Its value is in the range 30-40 mV, and only at higher frequencies, values of 80 mV were monitored. For 45 Hz and 55 Hz measuring points, due to depolarization or inability of the dipoles of both materials to follow the changes of the mechanical load with the same speed, the PZT layer cannot produce a charge. This behavior is probably due to the increased inner capacitance of the double layered structure.

In figure 13 some of the signals of the generated piezoelectric voltage, obtained from the structures are presented.

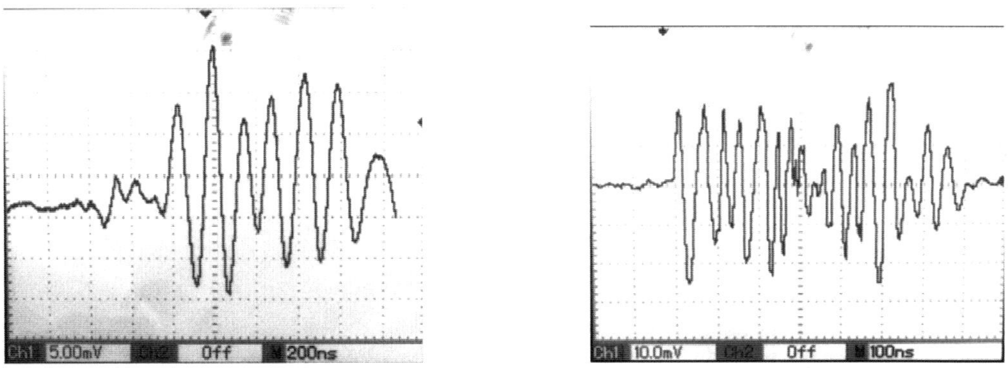

(a) (b)

Figure 13. Output voltage signals of the double-layer structure with PZT and ZnO layers for different frequencies [24].

For the second type of double layer structure, the first deposited material was ZnO and the second one - PZT. Like in the previous case, the value of generated piezoelectric voltage is almost a half of that for the single-layer structures. Here, at 23 Hz and 33 Hz, the voltage is around 40 mV and at 45 Hz and 55 Hz some values of 80-100 mV were rarely observed. Some signals, of the piezoelectric voltage, which is generated by this double-layer structure, are shown in figure 14.

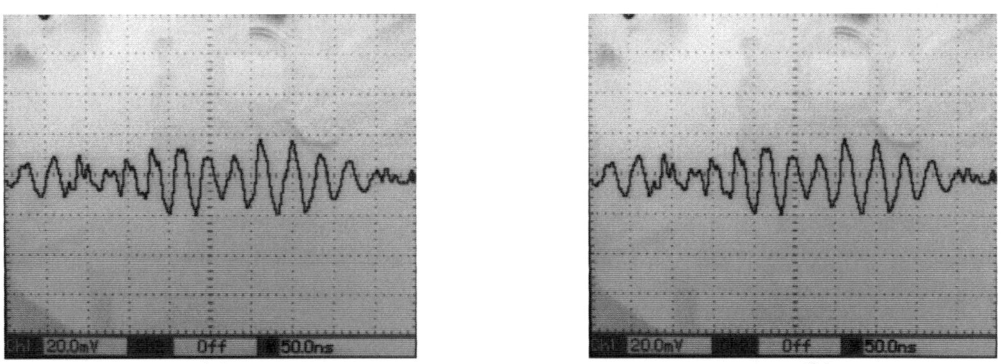

(a) (b)

Figure 14. Output voltage signals of the double-layer structure with ZnO and PZT [24].

After considering the generated piezoelectric voltage signals of all structures, it could be concluded that both two-layer structures have a lower output voltage. They actually generate a higher charge, but the inner capacitance is bigger and, as a result, the output measured voltage is lower. For one-layer structures, more interesting is the fact that the 60 nm ZnO layer has almost the same voltage values as the voltage, obtained by the 200 nm PZT layer. This could be a consequence of the piezoelectric coefficients of the layers. Another fact is that, the value of the output voltage, at 23 Hz and 33 Hz, for single-layer samples, is lower but for two-layer sample is higher. At higher frequencies, it is just in opposite, but future work should be done to clarify the processes occurred.

5. Conclusion

Based on the experiments it could be said that the PZT and ZnO based thin film devices can be successfully applied as piezoelectric stress sensors, energy harvesters and piezoelectric transformers. Piezoelectric sensor structures, using sputtered PZT layers, operate in different ranges of the applied stress, depending on the electrodes configuration. Using of flexible PET substrate provides higher elasticity of the piezostructure, which increases the voltage generation, even at lower mechanical loading and makes the energy harvesting devices more sensitive and effective. After considering the generated piezoelectric voltage signals of all structures, it could be concluded that, two-layer structure have a lower output voltage and can be used as piezoelectric decreasing transformers.

Acknowledgements: This paper was presented at INERA Conference „Vapor Phase Technologies for Metal Oxide and Carbon Nanostructures“, 5-9 July, 2016, Velingrad, Bulgaria. The Conference is part of the Program of INERA REGPOT Project of Institute of Solid State Physics, Bulgarian Academy of Sciences.

6. References

[1] Williams C and Yates R 1995 Analysis of a Micro-Electric Generator for Microsystems *IX International Conference Solid-State Sensors and Actuators and Eurosensors* (*Stockholm, Sweden, 25–29 June 1995*) pp 369-372
[2] Vullers R, Van Schaijk R, Doms I, Van Hoof C and Mertens R 2009 *Solid State El*ectron **53** 684
[3] Zribi A and Fortin J 2009 *Functional Thin Films and Nanostructures for Sensors - Synthesis, Physics and Applications* (Springer Science & Business Media)
[4] Trolier-McKinstry S and Muralt P 2004 *J. Electroceram.***12** 7
[5] Safari A and Akdogan E K 2008 *Piezoelectric and acoustic materials for transducer application* (Springer).
[6] Kolev G, Aleksandrova M, Videkov V, Denishev K and Truhchev P 2012 Piezoelectric MEMS Stress Sensor with Thin Lead Zirconate Titanate (PZT) Layer *20th Telecommunications Forum TELFOR (Belgrade, Serbia, 20-22 Nov. 2012)* pp 991-993
[7] http://www.nanomotion.com/piezo-ceramic-motor-technology/piezoelectric-effect/
[8] Vucheva Y, Kolev G, Aleksandrova M and Denishev K 2014 Determining the Piezoelectric Properties of Two-Layer Structures Based on Lead Zirconium Titanate and Zinc Oxide Thin Films in Dynamic Mode *16th International workshop on Nanoscience and Nanotechnology, NANO'2014 (Sofia, Bulgaria 07-08 November 2014)* p 73
[9] http://www.piezo.com/tech4history.html
[10] Dutsolova Y, Kolev G, Aleksandrova M, Andreev S, Dobrikov G and Denishev K 2013 Patterning of piezoelectric thin films from PZT and ZnO for application in MEMS devices *XI National Scientific and Practical Conference for Young Scientists (Sofia, Bulgaria, 25–26 April 2013) pp 76-79*
[11] https://www.memsnet.org/material/leadzirconatetitanatepzt/?keywords=pzt
[12] Kolev G. 2012 *Annual Journal of Electronics* **6** 88
[13] Johnson R 2005 *Characterization of piezoelectric ZnO thin films and the fabrication of piezoelectric micro-cantilevers* (Iowa State University)

[14] Kolev G., Denishev K, Aleksandrova M and Dutsolova Y 2013 Investigation of Thin PZT and ZnO Piezoelectric Layers in Dynamic Mode, for Application in *MEMS XLVIII International Scientific Conference on Information, Communication and Energy Systems and Technology, "ICEST 2013" (Ohrid, Macedonia 26-29 June 2013)* pp 691-69
[15] Muralt P 2008 *J. Am. Ceram. Soc.* **91** 1385
[16] Li C, Jang B, Wang X, Li M, Su L and Li X 2011 *Appl. Surf. Sci.* **257** 5998-6003
[17] Muthukumarana S and Gopalakrishnan R 2012 *Optic. Mater.* **34** 1946
[18] Verma A, Kar M and Agnihotry S 2007 *Solar Energy Mater. Solar Cells* **91** 1305
[19] Pakizeh E, Hosseini S, Kompany A and Ghasemifard M 2010 *Inter. J. Nanosci.* **3** 193
[20] Pontes F , Leit E, Nuns M, Pontes D, Longo E, Magnani R, Pizani P and Varela J 2004 *J. European Ceramic Soc.* **24** 2969
[21] Zak A and Majid W 2010 *Ceram. Inter.* **36** 1905
[22] Kumari N, Singh A and Mandal J 2013 *Inter. Refer. J. Engin. Sci.* **2** 48
[23] Kolev G, Aleksandrova M, Vucheva Y and Denishev K 2014 *J. Phys.: Conf. Series* **559** 012015 doi:10.1088/1742-6596/559/1/012015
[24] Vucheva Y, Kolev G., Aleksandrova M and Denishev K 2015 Comparison of the Piezoelectric Properties of Single-Layer and Bilayer Structures with Thin Films of PZT and ZnO in Dynamic Mode *L Int. Sc. Conf. on Information, Communication and Energy Systems and Technologies ICEST 2015 (Sofia, Bulgaria, 24-26 June 2015)*

Journal of Physics: Conference Series **764** (2016) 012004 doi:10.1088/1742-6596/764/1/012004

Atomic layer deposition of ZnO:Al on PAA substrates

B S Blagoev[1, 5], E Vlakhov[1], V Videkov[2], B Tzaneva[2], G Łuka[3], B S Witkowski[3], P Terziyska[1], J Leclercq[1], T A Krajewski[3], E Guziewicz[3], D Z Dimitrov[1, 4], V B Mehandzhiev[1] and P Sveshtarov[1]

[1] Institute of Solid State Physics, Bulgarian Academy of Sciences, 72 Tzarigradsko Chaussee Blvd, 1784 Sofia, Bulgaria
[2] Technical University – Sofia, 8 Kl. Ohridski Blvd, 1000 Sofia, Bulgaria
[3] Institute of Physics, Polish Academy of Sciences, Al. Lotnikow 32/46, 02-668 Warsaw, Poland
[4] Institute of Optical Materials and Technologies, Bulgarian Academy of Sciences, Acad. G. Bonchev Str., bl.109, 1113 Sofia, Bulgaria

E-mail: blago@issp.bas.bg, blago_sb@yahoo.com

Abstract. In this work the ZnO:Al films of different thickness are grown on the Porous Anodic Alumina (PAA) and p-Si (100) substrates by Atomic Layer Deposition. The ZnO:Al films thicknesses are chosen appropriately in order to obtain complete filled pores as well as pores with a thin covering on the surface. The obtained structures are investigated with spectroscopic ellipsometry and Scanning Electron Microscopy (SEM) techniques.

1. Introduction

In last decades the interest in nanomaterials increases rapidly due to their profound implications for many sectors of the electronics [1]. Two principal factors cause the properties of nanomaterials differ significantly from other materials: increased relative surface area, and quantum effects. The unusual properties of nanomaterials give their rich variety of applications: nanoelectronics, optoelectronics, spintronics, sunscreens and cosmetics, paints, displays, batteries, catalysis, medicine, sensors, food, construction, agriculture, energy and many others [1]. Therefore, a lot of companies and scientific laboratories all over the world are involved in development and investigation of nanomaterials. One of the easiest techniques for obtaining of cost-less templates for construction and designing different nanostructures is Porous Anodic Alumina (PAA) or Anodic Aluminum Oxide (AAO). PAA is a self-organized material with honeycomb-like structure formed by high density arrays of uniform and parallel nanopores with controlled diameter of the pores (from a few nm to several hundreds nanometers) formed by electrochemical oxidation (anodization) of aluminum in different electrolytes [2–5]. The deposition of nanofilms on 3D substrates like Porous Anodic Alumina (PAA) is of great interest from fundamental point of view as well as for practical applications. The best technique for covering such complicated substrates is the Atomic Layer Deposition (ALD) method because of its perfect 3D conformability, uniform covering on ultra-high aspect ratio (>2000:1) features, large area thickness uniformity, atomically flat and smooth coatings and possibility to control the thickness at the nanometer scale. Depending on the PAA substrate properties (pores dimensions, distance between pores, one or double side opened structures), deposition conditions (type of precursor, pulse duration, repetition, substrate temperature, etc.), chemical and thermal treatment different nanostructures could be obtained such as nanorods, nanotubes and nanodots. These nanostructures could consist of single or multi shell constructions in which each shell could be formed from different material (conducting, insulating or magnetic) which gives a rich variety of possibilities. Such structures could be used as magnetic, gas, chemical and biological sensors, as well as for catalytic converters, medical applications, Solid Oxide Fuel Cells (SOFS), supercapacitors, batteries, magnetic, nanoelectronic, spintronic and magnonic devices.

In this work the ZnO:Al films of different thickness were grown on the PAA substrates by Atomic Layer Deposition. The thicknesses were chosen appropriately to obtain completely filled pores and pores with a thin covering on the surface. The structures were investigated by plan- view and cross-section Scanning Electron Microscopy (SEM).

2. Experimental

AAO templates were prepared by two-step anodization of 38 μm 99.9995% Al foil (Alfa Aesar). The first anodization of the Al plate was performed at a constant voltage of 40 V in 4% oxalic acid at 18 °C for 30 min after degreasing of the Al plate in 40 g/L NaOH followed by electropolishing for 60 and 30 seconds in specimens B1601-3 and B1601-4, respectively. Then the anodic oxide was removed in a solution of phosphoric and chromic acids at 80 °C for 70 min. The second anodization then was carried out to the complete transformation of Al to Al_2O_3. Furthermore, a pore widening step was done for samples B1601-1 and B1601-2 in the same electrolyte (4% $(COOH)_2$), 40 °C, 40 min.

Atomic Layer Deposition (ALD) was used to deposit ZnO:Al films of different thickness. Thermal ALD with a Beneq TFS-200 reactor at 200 °C in nitrogen flow of 600 sccm was used. To obtain the ZnO:Al films the TMA (trimethylaluminium) with H_2O and DEZ (diethylzinc) with H_2O precursors at 1:24 ratio for Al_2O_3 and ZnO sublayers respectively were used. The duration of the precursor pulses and purging were 200 ms and 2 s respectively. The covering of PAA templates with different thicknesses was obtained by varying the repetition of full cycles ($ZnO:Al_2O_3$= 24:1 denotes one full cycle). The repetition of full $ZnO:Al_2O_3$ cycles 56; 4; and 2 times were done. In all processes the p-Si(111) substrate as a reference sample was loaded together with PAA templates. The samples were denoted as 28, 29 and 31 for 56, 2 and 4 full cycles repetition, respectively.

Ellipsometric measurements for Si reference substrates were performed using a Woollam M2000D rotating compensator spectroscopic ellipsometer with a wavelength range from 193 to 1000 nm. The film thickness and the optical constants were determined by fitting the experimental Ψ and Δ data.

SEM (Scanning Electron Microscope) investigations were carried out with the Hitachi SU-70 equipment using 15 kV of electron beam accelerating voltage and upper SE (Secondary Electron) detector, sensitive to the surface topography.

3. Results and discussion

3.1. Morphology of porous anodic alumina templates

Four different double–side Porous Anodic Alumina (PAA) templates (see figure 1) were obtained by a two–step anodization in the way described above. The main parameters of these templates are summarized in table 1. In our previous SEM investigations [6] it was determined that at similar experimental conditions (as in samples B1601-3 and B1601-4) the diameters of the pores were about 30-40 nm and grew up to 60-70 nm after their widening (samples B1601-1and B1601-2). Polishing was used to reduce the Al foil thickness and thereby obtain anodic oxide of different thickness. The final thickness of PAA templates was measured using optical microscopy technique (figure 2).

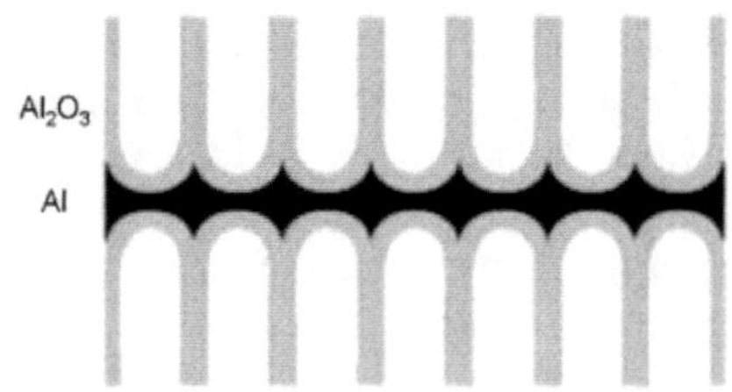

Figure 1. Sketch of double side PAA.

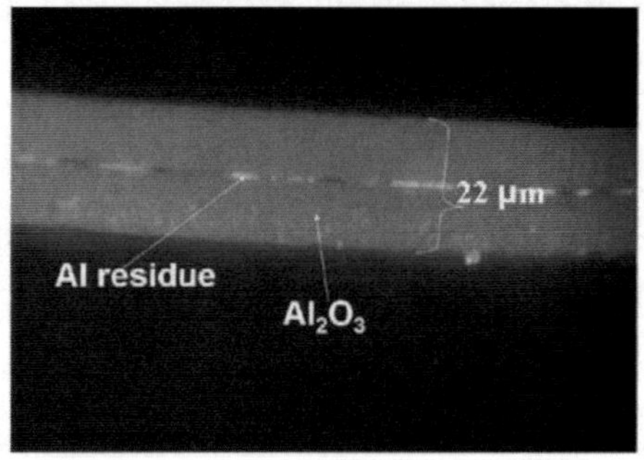

Figure 2. Optical microscopic image (cross-section) of the B1601-1 sample.

Table 1. Parameters of PAA templates.

	Pore diameters, nm	Total thickness of PAA templates, µm
B1601-1	60-70	22
B1601-2	60-70	30.5
B1601-3	30-40	17
B1601-4	30-40	22

3.2. Parameters of ALD thin films deposition

Thermal ALD depositions of the ZnO:Al thin films of different thickness on PAA templates and p-Si(111) substrates were done. In order to obtain conformal films, the Al_2O_3 monolayer was grown first (1 cycle) from the TMA and H_2O precursors. Then, 24 monolayers of ZnO were deposited (24 cycles) from the DEZ and H_2O precursors. The full cycle (ZnO:Al 24:1) was repeated 56, 4 and 2 times to obtain different thickness. The deposition rates of pure ZnO and Al_2O_3 sublayers were obtained from thick films (1010 and 880 cycles respectively) grown on Si substrates at 200 °C and the thicknesses were extracted from the Ellipsometric measurements [7]. Deposition rate for ZnO and Al_2O_3 films are 2 and 1.14 Å/cycle respectively. Therefore, for full cycle (ZnO:Al 24:1) with repetition 56, 4 and 2 the obtained thicknesses correspond to 275, 19.7 and 9.8 nm. Moreover, the volume percentage of Al_2O_3 content in ZnO:Al films evaluated from the formula of $\alpha = \frac{mA}{mA + nB}$, given in [7] should be about 2.3 %, where the m and n are the numbers of TMA and DEZ cycles and A and B are the growth rates of Al_2O_3 and ZnO, respectively. It should be noted that in thinner films (with 2 and 4 repetition of full cycles) the growth rate is supposed to be lower due to the interphase interactions and surface initialization for conformal deposition. As the repetitions number increases the surface conformality improves and growth rate becomes more uniform. These factors also should lead to the Al content decrease in the ZnO:Al films for their lower thicknesses.

3.3. Ellipsometric study

The spectroscopic ellipsometry data of Ψ and Δ were taken in the wavelength range from 193 to 1000 nm on ZnO:Al (24:1) films grown on p-Si(111) with repetition of full cycles 56, 4, and 2, respectively. The experimental Ψ and Δ were analyzed using a three-layer model consisting of a silicon substrate with SiO_2 native oxide as a first layer, a ZnO layer as a second layer and a roughness layer as a third layer. For the Si substrate and the native oxide, the data from the database of CompleteEASE Woollam data analysis software was used. The ZnO:Al layer was modeled using a PSemi-M0 and two Gaussian oscillators. The roughness layer for all samples is modeled by Bruggeman's EMA (Effective

Medium Approximation) of 50 % voids and 50 % bulk material [8]. The thicknesses of the ZnO:Al layers on Si substrates as determined from the fits are as follows: 262.46 nm for sample 28, 8.11 nm for sample 29 and 16.54 nm for sample 31. These results are in agreement with the thicknesses obtained from the growth rate discussed above.

We estimated the optical bandgap E_g of the ZnO:Al from the equation $\alpha = \frac{(h\nu - E_g)^{1/2}}{h\nu}$, by plotting the absorption coefficient squared $(\alpha h\nu)^2$ versus photon energy, as shown on figure 3. The dispersion of the absorption coefficient α was determined from the dispersion of the extinction coefficient $\kappa(\lambda)$ using the equation $\alpha = 4\pi\kappa/\lambda$. The extinction coefficient κ was estimated by fitting the experimental data Ψ and Δ.

Figure 3 shows a comparison of the squared absorption coefficient $(\alpha h\nu)^2$ versus photon energy for 207 nm thick undoped ZnO layer and a 262.46 nm thick ZnO:Al (sample #28) layer of similar thicknesses. The estimated optical bandgap of the aluminum doped zinc oxide layer is about 210 meV larger than the optical bandgap of the undoped ZnO layer. This is supposed to be due to the Burstein-Moss effect. The samples #29 and #31 were omitted because of the band gap expansion due to the quantum confinement effect as described in (Ref. [9, 10]).

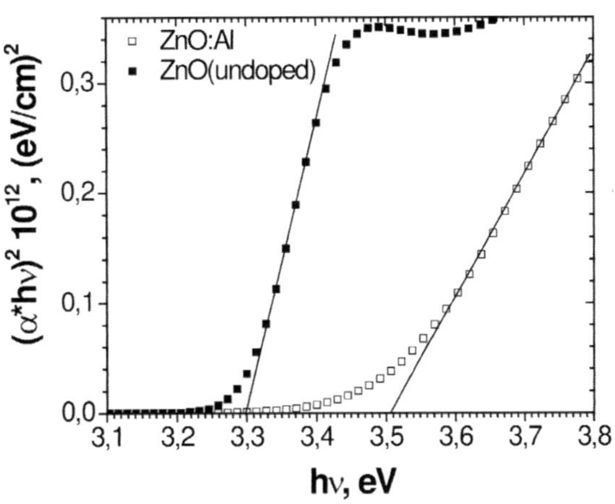

Figure 3. Square of absorption coefficient $(\alpha h\nu)^2$ for undoped and Al-doped ZnO layers *vs.* photon energy. Solid lines show linear fits to the experimental data. The intercept at $(\alpha h\nu)^2 = 0$ shows the bandgap of ZnO.

3.4. SEM observations

The SEM images show an extremely conformal coverage of the PAA matrix by ZnO:Al layer deposited by ALD (figures 4 and 5). This layer fills the pores even up to a depth of several micrometers (which is unattainable with other deposition methods). The ZnO:Al layer is polycrystalline (as shown on SEM images) and pore-free (dense material, which is characteristic for the materials obtained by ALD).

On the figure 4, plan- view and cross- section SEM images of ZnO:Al films on PAA templates for two different thicknesses corresponding to full cycle repetition of 4 and 56 are presented.

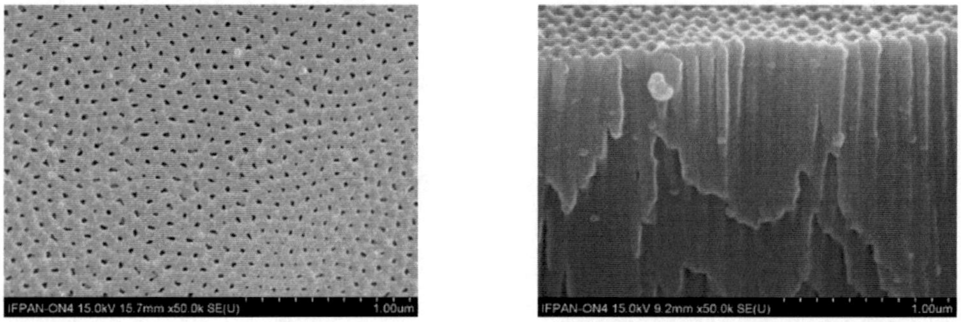

Figure 4. SEM images of ZnO:Al with 4 times repetition on PAA template (a) top- view and (b) cross- section of sample B1601-2 2/31

At 4 repetitions (around 16 nm ZnO:Al on Si substrate according to the ellipsometry data) of full cycle very thin film of ZnO:Al covers the walls of pores and decreases the pores diameters. For sample B1601-2 2/31 (with 60-70 nm pores) the pores diameters decrease to 30-40 nm (figure 4). At lower magnification the typical honeycomb-like structure of PAA template can be observed (figure 4a). On higher magnification (not shown here) it should be seen the grains 20×10 nm. The higher repetition (56 full cycles) fill up the pores and dense ZnO:Al films (~270 nm) start to grown on the top (figure 5) with the typical for ZnO sticks-like structure (200×20 nm elongated grains). The obtained thickness from SEM images 270 nm is a little bit higher than the thickness of 260 nm obtained by ellipsometric measurements on a Si substrate because of the higher deposition rate on PAA due to already saturated the Al_2O_3 surface.

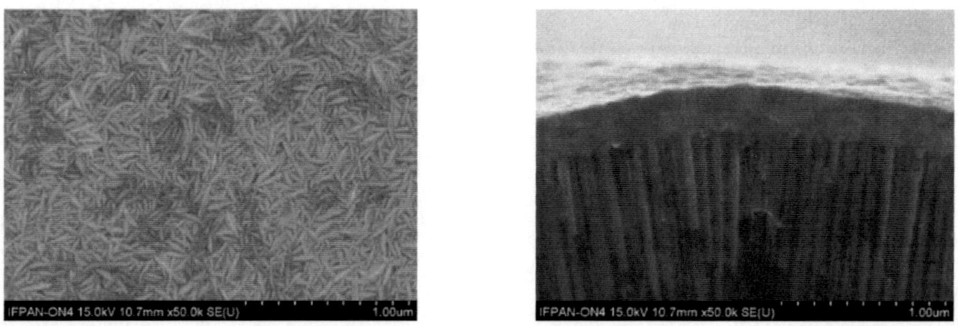

Figure 5. SEM images of ZnO:Al with 56 time repetition on PAA template (a) top -view and (b) cross- section of sample B1601-2 3/28

4. Conclusions

ZnO:Al (24:1) films with different thicknesses were grown by Atomic Layer Deposition on Porous Anodic Alumina templates and p-Si(100) substrates. An extremely conformal coverage of the PAA templates with a ZnO:Al layer is obtained as shown by SEM analysis. The layers fill the pores even up to a depth of several micrometers which is unattainable with other deposition methods. The deposited ZnO:Al layer is polycrystalline and pore-free dense material. Ellipsometry analyses on Si substrates confirm the thickness of ZnO:Al films obtained from the SEM. It was found that the estimated optical bandgap from the Ellipsometric measurements of the thicker Al doped ZnO layer is about 170 meV larger than the optical bandgap of the undoped ZnO layer due to the Burstein-Moss effect.

Acknowledgments
This work was supported by the EU 7th FP REGPOT project INERA (GA No 316309).

References

[1] Arivalagan K, Ravichandran S, Rangasamy K and Karthikeyan E 2011 *Int. J. Chem. Tech. Research* **3** 534
[2] Ono S, Saito M and Asoh H 2005 *Electrochimica Acta* **51** 827
[3] Garcia-Vergara S J, Skeldon P, Thompson G E and Habazaki H 2006 *Electrochimica Acta* **52** 681
[4] Garcia-Vergara S J, Skeldon P, Thompson G E and Habazaki H 2007 *Thin Solid Films* **515** 5418
[5] Thompson G E 1997 *Thin Solid Films* **297** 192
[6] Tzaneva B R, Naydenov A I, Todorova S Z, Videkov V H, Milusheva V S and Stefanov P K 2016 *Electrochimica Acta* **191** 192
[7] Blagoev B S, Dimitrov D Z, Mehandzhiev V B, Kovacheva D, Terziyska P, Pavlic J, Lovchinov K, Mateev E, Leclercq J and Sveshtarov P 2016 *J. Phys.: Conf. Series* **700** 012040
[8] CompleteEASE® Software Manual, J.A.Woollam Co. Inc.
[9] Li X D, Chen T P, Liu P and Leong K C 2014 *Journal of Applied Physics* **115** 103512
[10] Li X D, Chen T P, Liu P, Liu Y and Leong K C 2013 *Optical Express* **21** 14138

Journal of Physics: Conference Series **764** (2016) 012005 doi:10.1088/1742-6596/764/1/012005

Core-shell carbon nanosphere-TiO_2 composite and hollow TiO_2 nanospheres prepared by atomic layer deposition

L P Bakos[1], N Justh[1], K Hernádi[2], G Kiss[2], B Réti[2], Z Erdélyi[3], B Parditka[3] and I M Szilágyi[1,4]

[1] Department of Inorganic and Analytical Chemistry, Budapest University of Technology and Economics, Hungary
[2] Department of Applied and Environmental Chemistry, University of Szeged, Hungary
[3] Department of Solid State Physics, University of Debrecen, Hungary
[4] MTA-BME Technical Analytical Chemistry Research Group, Hungary

E-mail: laszlobakos@hotmail.com
justh.nora@gmail.com
imre.szilagyi@mail.bme.hu

Abstract. Core-shell carbon-TiO_2 composite and hollow TiO_2 nanospheres were prepared using carbon nanospheres as hard-templates, coating them with TiO_2 using atomic layer deposition, and subsequent burning out of the carbon cores. The bare carbon, the composite carbon-TiO_2 and the hollow TiO_2 nanospheres were characterized with TG/DTA-MS, FTIR, XRD and SEM-EDX.

1. Introduction

Photocatalysis using solar energy has been acknowledged as an environment friendly approach to address environmental problems, such as pollutant degradation, heavy metal cation reduction and treatment of wastewater [1]. Titanium oxide materials have great potential as photocatalysts in various reactions due to their chemical stability, non-toxicity, and high reactivity [2,3]. However, the fast recombination of photo-excited charge carriers leads to their short lifetime, which interferes with achieving maximum activity of TiO_2 photocatalysts, so it is desirable to use a co-catalyst to synthesize economically feasible TiO_2-based photocatalysts with improved charge separation and low electron–hole recombination rates. To fulfill this, several carbon materials have been utilized to synthesize carbon–TiO_2 composite photocatalysts [4,5]. Carbon-based nanomaterials are very attractive, due to their high surface area, good thermal and electrical conductivity, mechanical as well as chemical stability, and they provide an ideal support for heterostructure elaboration. Spherically shaped carbon materials have been known for decades and their properties have been exploited in many areas (e.g. tyres, batteries, printer inks) [6,7]. Inorganic hollow spheres, such as ZnS, WO_3, TiO_2 and SiO_2, have special optical, optoelectronic, magnetic, electrical, thermal, electrochemical, photoelectrochemical and catalytic properties, and have been investigated due to their potential application in catalysts, lithium-ion batteries, biomedical materials and gas sensors [8,9].

There are several preparation methods for creating carbon nanospheres as cores for carbon TiO_2 composites, such as laser ablation, chemical vapor deposition and hydrothermal synthesis [7]. Atomic layer deposition (ALD) allows the coating of surface of nanostructures in a conformal and homogeneous way, with precise control of the thickness of the deposited film at the nanoscale [6,10–14]; hence, it is an outstanding possible tool to coat carbon nanospheres.

In this study, we prepared carbon nanospheres using hydrothermal synthesis, and coated them with TiO_2 layer using ALD to obtain a composite material. In order to get the hollow TiO_2 nanospheres, we burnt out the carbon cores of the composite. We used thermogravimetry and differential thermal analysis coupled with mass spectrometry (TG/DTA-MS), Fourier transform infrared spectroscopy (FTIR), powder X-ray diffraction (XRD) and scanning electron microscopy with energy dispersive X-Ray analysis (SEM-EDX) to characterize the as-prepared materials.

2. Materials and characterization

2.1. Preparation of the carbon nanospheres

To prepare the carbon nanospheres, the hydrothermal method was used [7]. A sucrose solution of 0.15 M concentration was placed into an autoclave with a volume of 275 ml. The pH was set to 12 using 0.194 M NaOH solution. The reaction went for 12 hours at 180 °C under autogenous pressure. Then the resulting product was washed with distilled water three times, suspended in 45 V/V % ethanol-water solution, centrifuged for 20 min at 4000 rpm. After that, the product was filtered, washed with warm distilled water, and placed in drying cabinet at 70 °C for overnight. The resulting product was fine black powder. This method was similar to the one described in [15].

2.2. Preparation of the composites

TiO_2 was deposited by atomic layer deposition on the carbon nanosphere powder using $TiCl_4$ and H_2O as precursors in a Beneq TFS-200-186 ALD thermal reactor at 1 mbar pressure. The pulse and purge times for one cycle were 0.3 s/3 s and 0.3 s/3 s for the $TiCl_4$ and H_2O precursors, respectively. The depositions were carried out at 250 °C, and 700 ALD cycles were used.

2.3. Characterization

TG/DTA measurements were conducted on a TA Instruments SDT 2960 simultaneous TG-DTA device in helium or air atmosphere (130 ml/min) using an open platinum crucible and 10 °C/min heating rate. EGA-MS (evolved gas analytical) curves were recorded by a Balzers Instruments Thermostar GSD 200T quadruple mass spectrometer (MS) coupled on-line to the TG/DTA instrument; the on-line coupling between the two parts was provided through a heated (T=200 °C), 100 % methyl deactivated fused silica capillary tube with inner diameter of 0.15 mm [16].

FTIR measurements were carried out between 4000 and 400 cm^{-1} on a Biorad Excalibur Series FTS 3000 infrared spectrometer. 300 mg KBr pellets were used, which contained 1.0 mg sample. 64 measurements were accumulated into one spectrum.

Powder XRD patterns were recorded on a PANanalytical X'Pert Pro MPD X-ray diffractometer using Cu Kα radiation.

SEM-EDX data were obtained by a JEOL JSM-5500LV scanning electron microscope and the samples were sputtered with an Au/Pd conductive layer for the imaging.

3. Results

3.1. TG/DTA-MS

Figure 1 shows the TG/DTA-MS analysis of pure carbon nanospheres in helium. First, the nanospheres lose a minimal amount of water (m/z=18), then the decomposition begins around 300 °C, which can be seen from the evolving gases of CO_2 (m/z=44) and cyclohexene (m/z=82). This ensured that the ALD depositions could be done safely at 250 °C without damaging the template.

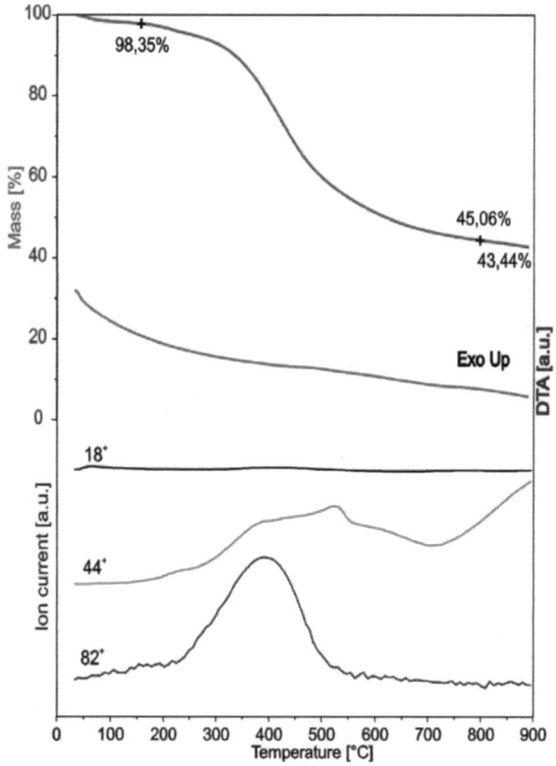

Figure 1. TG/DTA-MS analysis of the carbon nanospheres in helium

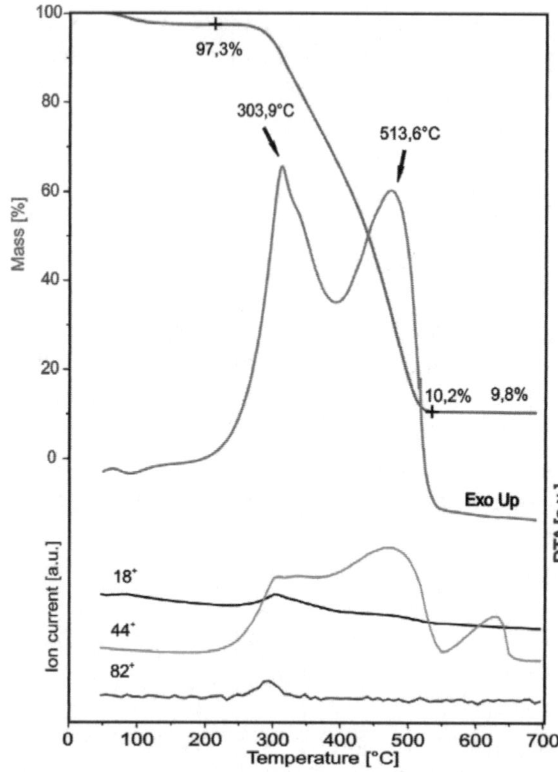

Figure 2. TG/DTA-MS analysis of the carbon nanosphere-TiO_2 composite in air

Figure 2 displays the thermal analysis of carbon nanosphere-TiO_2 composite in air. In summary, the carbon core burnt out and only the hollow TiO_2 nanospheres remained in the platinum crucible. The molecule-ion of CO_2 (44^+) was dominant in the mass spectrum of the evolved gaseous products, and unlike the analysis in helium, the cyclohexene was much less abundant because of its oxidation. In the case of both EGA analyses, the intensity of the ion current curve of the CO_2 increased at higher temperatures (above 600 °C), owing to the decomposition of the polymer coating of capillary between the TG/DTA and the MS [4,9,17,18].

3.2. FTIR

In the FTIR spectra (figure 3) at 1600 cm^{-1} the characteristic peak of the carbon-carbon double bond can be seen, and at 2850 cm^{-1} the characteristic peak of the carbon-hydrogen bond is visible, both of which disappeared after burning out of the carbon nanosphere template [7]. At 1700 cm^{-1} and 3400 cm^{-1} the characteristic peaks of the OH-group are present [19]. The lattice vibration bands of the TiO_2 are under 500 cm^{-1}, which can only be seen at the spectrum of the hollow TiO_2 shell, because there was only small amount of TiO_2 on the carbon nanosphere-TiO_2 composite [20].

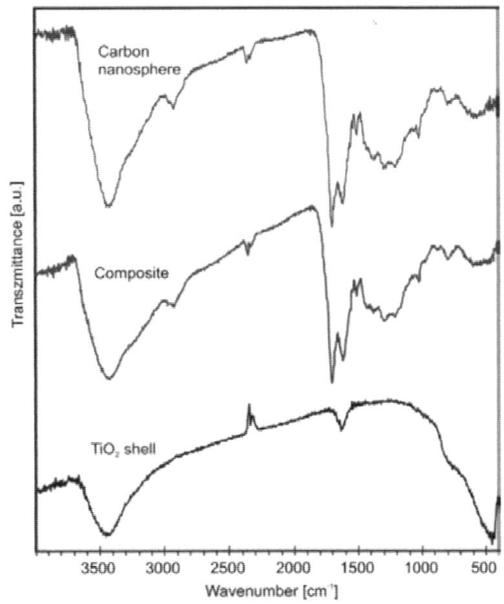

Figure 3. FTIR spectra of the samples

3.3. XRD

From the powder XRD diffractograms (figure 4) it can be seen that both the carbon nanospheres and the carbon nanosphere-TiO_2 composite are amorphous; in the latter the peaks of TiO_2 are not recognizable due to the low amount of it. The hollow TiO_2 shell is crystalline and is identified as anatase (ICDD 01-075-2546).

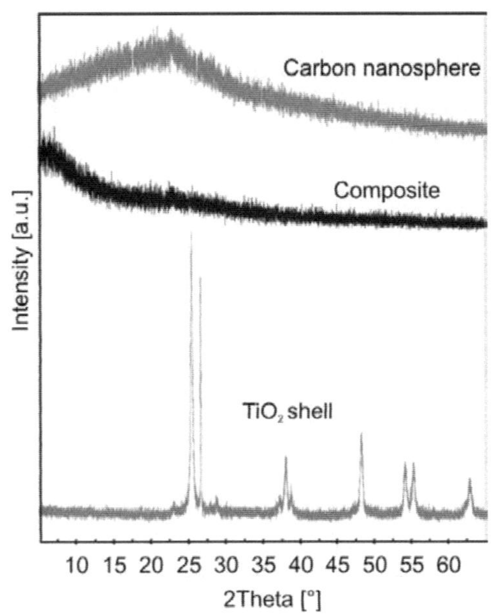

Figure 4. Powder XRD patterns of the samples

3.4. SEM-EDX

From the SEM images (figure 5) the spherical shape of carbon particles can be seen, which was maintained both after the ALD reaction and the burning out of the carbon core as well. The steps of the synthesis could be followed by the EDX results as well (table 1). The atomic % of the C decreases after the ALD process, while the O and Ti increases. The burning out of the carbon core was successful, only 12.7 atomic % carbon remained. The Cl is residue from the $TiCl_4$ precursor of the ALD deposition.

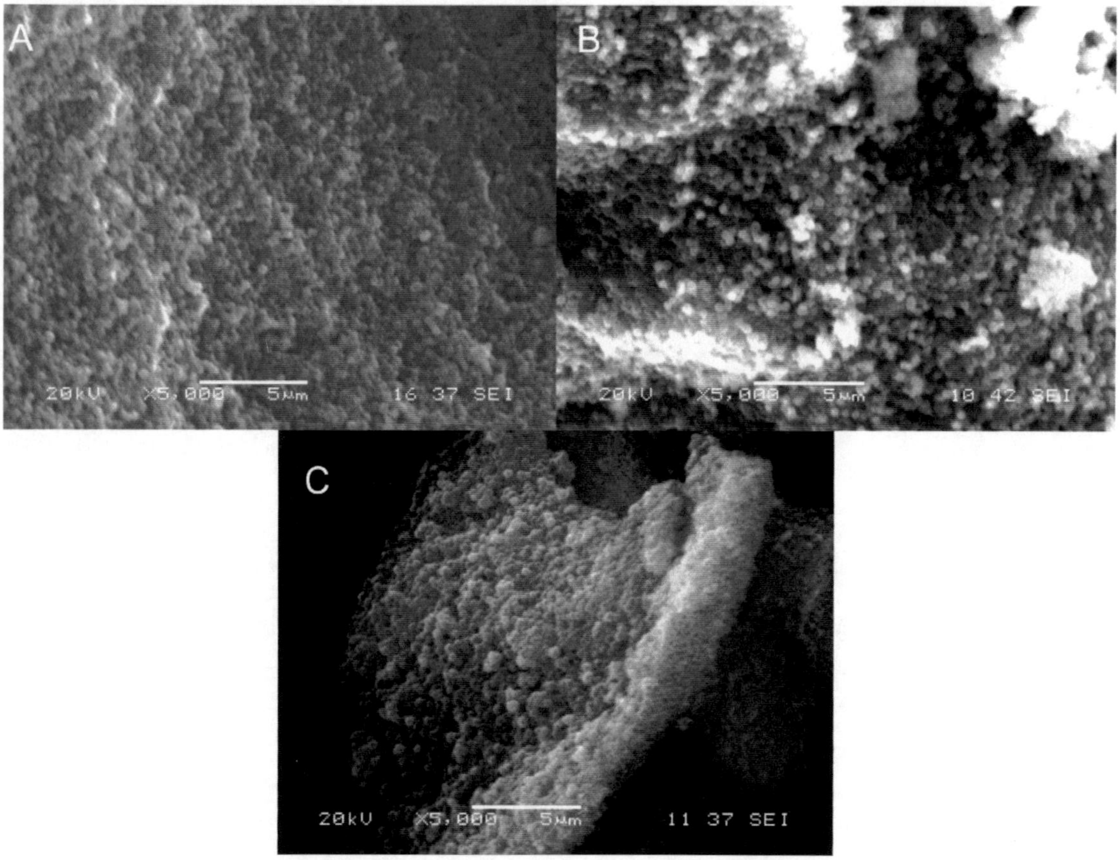

Figure 5. SEM images of (A) the carbon nanospheres; (B) the carbon nanosphere-TiO_2 composite material and (C) the hollow TiO_2 shells

Table 1. Composition of the samples from the EDX spectra

	C	O	Ti	Cl
		atomic %		
Carbon nanospheres	77.3	22.7	0.0	0.0
Carbon nanosphere-TiO_2 composite	62.0	27.8	10.1	0.1
Hollow TiO_2 nanospheres	12.7	69.4	17.8	0.1

4. Conclusions

We successfully prepared carbon nanospheres using the hydrothermal method. Consecutively, we deposited TiO_2 film on the carbon nanospheres by atomic layer deposition and then burnt out the carbon cores to get hollow TiO_2 nanospheres. The ALD deposition and the formation of the hollow TiO_2 shells were followed by TG/DTA-MS, FTIR, XRD and SEM-EDX techniques. The TG/DTA-MS measurement of the carbon nanospheres in helium confirmed that the ALD process could be done at 250 °C without damaging the template. Based on the results of TG/DTA-MS of the carbon/TiO_2 composite in air, the carbon cores burnt out between 200 and 600 °C. The FTIR and EDX measurements proved the presence of the oxide in the composite and the effectiveness of the following annealing. The XRD showed that the hollow TiO_2 nanospheres were crystalline (anatase). The SEM images revealed that the spherical shape was maintained through the whole process. The carbon-nanosphere-TiO_2 composite and the hollow TiO_2 nanospheres can have great potential in the field of photocatalysis using solar energy.

5. References

[1] Yu C, Zhou W, Liu H, Liu Y and Dionysiou D D 2016 Design and fabrication of microsphere photocatalysts for environmental purification and energy conversion *Chem. Eng. J.* **287** 117–129

[2] Schneider J, Matsuoka M, Takeuchi M, Zhang J, Horiuchi Y, Anpo M and Bahnemann D W 2014 Understanding TiO2 Photocatalysis: Mechanisms and Materials Chem. Rev. **114** 9919–9986

[3] Nakata K and Fujishima A 2012 TiO2 photocatalysis: Design and applications *J. Photochem. Photobiol. C Photochem. Rev.* **13** 169–189 [4] Bong J, Liu H, Jae Y, Dahl M, Yu H, Zaera F and Yin Y 2016 Tailored synthesis of C @ TiO_2 yolk – shell nanostructures for highly efficient photocatalysis *Catal. Today* **264** 261–269

[5] Leary R and Westwood A 2011 Carbonaceous nanomaterials for the enhancement of TiO_2 photocatalysis *Carbon* **49** 741–772

[6] Marichy C and Pinna N 2013 Carbon-nanostructures coated / decorated by atomic layer deposition : Growth and applications *Coord. Chem. Rev.* **257** 3232–3253

[7] Deshmukh A A, Mhlanga S D and Coville N J 2010 Carbon spheres *Mater. Sci. Eng. R Reports* **70** 1–28

[8] Hu J, Chen M, Fang X and Wu L 2011 Fabrication and application of inorganic hollow spheres *Chem. Soc. Rev. Chem. Soc. Rev* **40** 5472–5491

[9] Song C, Yu W, Zhao B, Zhang H, Tang C, Sun K, Wu X, Dong L and Chen Y 2009 Efficient fabrication and photocatalytic properties of TiO2 hollow spheres *Catal. Commun.* **10** 650–654

[10] Szilágyi I M and Nagy D 2014 Review on one-dimensional nanostructures prepared by electrospinning and atomic layer deposition *J. Phys. Conf. Ser.* **559** 012010

[11] Boyadjiev S I, Santos G dos L, Szűcs J and Szilágyi I M 2016 Preparation and characterization of WO3 nanoparticles, WO3/TiO2 core/shell nanocomposites and PEDOT:PSS/WO_3 composite thin films for photocatalytic and electrochromic applications *AIP Conf. Proc.* **1722** 1–4

[12] Boyadjiev S, Georgieva V, Vergov L, Baji Z, Gáber F and Szilágyi I M 2014 Gas sensing properties of very thin TiO2 films prepared by atomic layer deposition (ALD) *J. Phys. Conf. Ser.* **559** 012013

[13] Szilágyi I M, Teucher G, Härkönen E, Färm E, Hatanpää T, Nikitin T, Khriachtchev L, Räsänen M, Ritala M and Leskelä M 2013 Programming nanostructured soft biological surfaces by atomic layer deposition *Nanotechnology* **24** 245701

[14] Szilágyi I M, Santala E, Heikkilä M, Pore V, Kemell M, Nikitin T, Teucher G, Firkala T, Khriachtchev L, Räsänen M, Ritala M and Leskelä M 2013 Photocatalytic properties of WO_3/TiO_2 core/shell nanofibers prepared by electrospinning and atomic layer deposition *Chemical Vapor Deposition* **19** 149–155

[15] Ao Y, Xu J, Fu D and Yuan C 2008 A simple method for the preparation of titania hollow sphere *Catal. Commun.* **9** 2574–2577

[16] Szilágyi I M, Madarász J, Pokol G, Hange F, Szalontai G, Varga-Josepovits K and Tóth A L 2009 The effect of K+ ion exchange on the structure and thermal reduction of hexagonal ammonium tungsten bronze *J. Therm. Anal. Calorim.* **97** 11–18

[17] Ristić I S, Barta Holló B, Budinski-Simendić J, Mészáros Szécsényi K, Cakić S, Szilágyi I M and Pokol G 2015 Synthesis of novel metal-containing epoxy polymers and their structural characterization by means of FT-IR and coupled TG/MS measurements J. *Therm. Anal. Calorim.* **119** 1011–1021

[18] Palmer H B, Voet A and Lahaye J 1968 Mass spectra of hydrocarbon byproducts in the formation of carbon blacks *Carbon* N. Y. **6** 65–70

[19] Lin Y, Jin J and Song M 2011 Preparation and characterisation of covalent polymer functionalized graphene oxide *J. Mater. Chem.* **21** 3455

[20] Adamczyk A and Długoń E 2012 The FTIR studies of gels and thin films of Al_2O_3–TiO_2 and Al_2O_3–TiO_2–SiO_2 systems *Spectrochim. Acta Part A Mol. Biomol. Spectrosc.* **89** 11–17

Journal of Physics: Conference Series **764** (2016) 012006 doi:10.1088/1742-6596/764/1/012006

Investigations on Poly (ethylene oxide) (PEO) – blend based solid polymer electrolytes for sodium ion batteries

H K Koduru,[1,2]* M T Iliev,[2]K K Kondamareddy,[4]D Karashanova,[3]T Vlakhov,[2]X-Z Zhao,[4] N Scaramuzza[1,5]

[1]Dipartimento di Fisica, Università degli Studi della Calabria, Via P. Bucci, Cubo 33B - 87036 Rende (CS) – Italy.
[2]GeorgiNadjakov Institute of Solid State Physics, Bulgarian Academy of Sciences, 72 TzarigradskoChaussee, Sofia 1784, Bulgaria.
[3]Institute of Optical Materials and Technologies Bulgarian Academy of Sciences Acad. Georgy Bonchev St., bl. 109 1113 Sofia, Bulgaria.
[4]School of Physics and Technology, Key Laboratory of Artificial Micro/Nano Structures of Ministry of Education, Wuhan University, Wuhan, 430072, China.[5]CNISM – Unità di Ricerca di Cosenza c/o Dipartimento di Fisica, Università degli
Studi della Calabria Via P. Bucci, Cubo 31C - 87036 Rende (CS) – Italy.

* E-mail: koduruharika@gmail.com

Abstract. Polymer blend electrolytes based on Polyethylene oxide (PEO) and polyvinyl pyrrolidone (PVP), complexed with $NaIO_4$ salt and Graphene oxide (GO) are investigated in the present report. The electrolytes are prepared by a facile solution cast technique. X-ray diffraction (XRD), Fourier transform infrared spectroscopy (FTIR), and scanning electron microscopy (SEM) are employed to study the influence of ion–polymer interactions on the micro structural properties of blend electrolytes. Measurements of electrical conductivity of the blend polymer complexes have been performed by using complex impedance spectroscopy in the frequency range 1 Hz - 1 MHz and within the temperature range 303 K – 343 K.A study on electrical conductivity properties of GO doped 'salt complexed electrolyte' systems is presented.

1. Introduction

Solid-state polymer electrolytes (SPEs) have attracted great attention by the researchers in the field of rechargeable metal-ion battery applications, to overcome a great deal of drawbacks of conventional liquid electrolytes especially related to safety issues originating from their high volatility and flammability [1].Currently, Li – ion battery technology has been widely employed in energy supply &storage systems due to their high energy density values compared to all other battery chemistries. However, the high cost, decrease of abundance, environmental impact and safety limitations concerning to Lithium materials impede their widespread implementation in future battery technologies [2]. Therefore, there is urgent necessity to search for alternative energy storage system technology capable of complementing the Li-ion battery technology. Amongst the accessible battery chemistries, sodium (Na) based rechargeable batteries (SIBs) have recently captured much attention because they are environmentally friendly, non-toxic, low cost and earth abundant materials.

Substantial research, development, and demonstration efforts are currently in progress to replicate the performance of well-established Li-ion batteries with sodium ion batteries [3].

Polyethylene oxide based polymer electrolytes have been employed extensively for battery applications since Armand et al. have demonstrated the feasibility of using PEO as a potential ion conductive electrolyte [4]. PEO is one of the widely investigated host polymer for synthesizing solid polymer electrolytes (SPEs) for alkali metal ion-conducting batteries due to its high electrochemical stability, good solvation power, complexation and ion dissociation abilities. Several researchers have reported about various sodium salts complexed PEO based polymer electrolytes for sodium ion battery applications [5]. It is worth to mention that, below the melting temperature ($\approx$ 330 K)pure PEO electrolytes consist of both semi-crystalline and amorphous regions, as well as intermediate regions at the crystalline/amorphous interphase. The semi-crystalline nature of PEO leads to both amorphous and crystalline phases at room temperature, which subsequently limits its ionic conductivity [6]. The suppression in the degree of crystallinity of the polymer chains improves their mobility, thereby leading to better ionic conduction. Therefore, ions hopping is believed to be predominant in amorphous rich PEO based electrolytes [7].The most straightforward approach to overcome this problem is to modify the PEO matrix in order to decrease its degree of crystallinity. One of the most promising alternate choices of enhancing the amorphous phase in PEO based electrolyte systems is blending of PEO with suitable higher amorphous polymer. Polyvinyl pyrrolidone (PVP) has been identified as a compatible partner to PEO which exhibits higher order amorphosity. PVP demonstrates high glass transition temperature, mechanical and thermal stabilities, provided the carbonyl group (C = O) in PVP enables to form different complexes of alkali metal ion salts at wide range of concentrations. In particular, reinforcement of nanometer sized materials with polymer electrolyte systems can modify their microstructural properties and enhance their electrochemical performances [8].

Blending of PEO based polymer electrolytes with nano dimensional carbon materials is a novel perspective approach to improve their electrochemical applications, especially for secondary batteries, ionic conductors, and supercapacitors energy storage systems. Graphene oxide (GO) has been recognized as one of the unique carbon materials because of its multiple oxygen-containing functional groups, which could enable it to be well dispersed in polar solvents and can improve the mobility of the mobile ions and lead to the improvement of the ion conductions in polymer electrolyte systems [9].In the present report, authors described about preparation of graphene oxide (GO) doped PEO/PVP blend system complexed with sodium periodate ($NaIO_4$) salt by using solution casting technique and discussed about microstructural and conductivity properties of prepared electrolytes.

2. Experimental

PEO of molecular weight 5 X 10^6 and PVP of molecular weight 3.6 X 10^5 were procured from Aldrich and employed without any further purification to prepare PEO/PVP solid state blend electrolytes. Sodium periodate salt ($NaIO_4$, Sigma Aldrich) and graphene oxide (GO of 2 mg/mL, dispersion in H_2O, Sigma Aldrich) were used as dopants and methanol (Aldrich) was used as a solvent.PEO/PVP blend electrolyte films of thickness 150 μm were prepared by following conventional solution cast technique. Appropriate amounts of PEO and PVP polymers were dissolved in methanol solution and followed mechanical stirring at room temperature for 10hours to get uniformity &miscibility of the polymers. In the meanwhile, different weight percentages (wt%) $NaIO_4$ salts were dissolved separately in methanol solution and added to pre-prepared viscous PEO/PVP polymer solutions to obtain $NaIO_4$ salt complexed 'PEO/PVP/$NaIO_4$ (x%) (x% = 0, 5, 7.5, 10)'polymer electrolytes. To study the influence of graphene oxide, GO solution was sonicated for 15 minutes and added appropriate amount to 'PEO/PVP/$NaIO_4$ (10 wt%)' electrolyte solution to prepare 'PEO/PVP/$NaIO_4$(10 wt%)/GO(0.4 wt%)' electrolytes. To harvest free-standing polymer electrolyte films, obtained viscous solutions were poured into polypropylene dishes and the solvent (methanol) was allowed to evaporate slowly at room temperature. All electrolyte films were vacuum dried at 45°C to remove traces of methanol

Journal of Physics: Conference Series **764** (2016) 012006 doi:10.1088/1742-6596/764/1/012006

solvent and kept in desiccators filled with silica gel desiccants for several hours before being characterized to avoid any traces of moisture.

The structural properties of blend electrolytes were analysed by recording X-ray diffraction patterns in the 2θ range of $10^{o} - 80^{o}$ with a constant step 0.02^{0}, counting time 35s/step on a Bruker D8 Advance diffractometer with Cu Kα radiation and LynxEye detector. To understand complexation of additives with the host polymer blend, FTIR measurements were carried out in nitrogen atmosphere to record transmittance spectra in the wavenumber range of 400–4000 cm^{-1} and with an optical resolution of 2 cm^{-1} using 'Bruker, model Vertex 70' spectrometer. Scanning electron microscopy (Philips 515 digitalized)measurements were performed to understand the modifications in surface morphological features of blend electrolytes as a function of complexation $NaIO_4$ salt and doping of GO. Ionic conductivity measurements were carried out by a. c. impedance measurements using Biologic potentiostat/galvanostat (SP – 200) from room temperature to 343 K in the frequency range of 1 Hz to 1 MHz. During the measurements, electrolyte films were sandwiched between two copper electrodes which were assembled in a temperature controlled furnace.

3. Results and Discussion

3.1. Microstructural properties

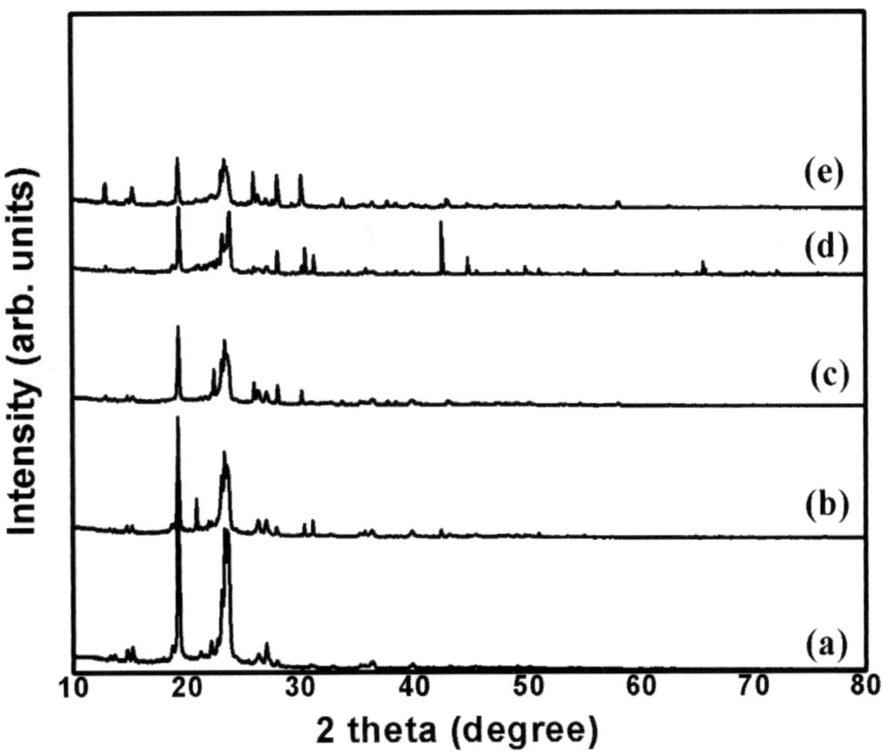

Figure 1. XRD spectra for pure, $NaIO_4$ complexed and GO doped PEO/PVP blend electrolytes: (a) PEO/PVP pure blend, (b) PEO/PVP/$NaIO_4$ (5 wt%), (c) PEO/PVP/$NaIO_4$ (7.5wt%), (d) PEO/PVP/$NaIO_4$ (10wt%), (e) PEO/PVP/$NaIO_4$ (10 wt%)/GO(0.4 wt%).

Figure 1 represents XRD patterns of pure blend (PEO/PVP) and its complexes with $NaIO_4$ salt at different concentrations and GO (0.4 wt%) doped 'PEO/PVP/$NaIO_4$ (10 wt%)' blend electrolytes. The diffraction pattern of pure blend (PEO/PVP) displays intensity diffraction peaks at $2\theta = 19.3^{0}$ and $2\theta = 23.4^{0}$ (figure 1(a)) corresponding to characteristic crystalline peaks of PEO, which originates from the ordering of polyether side chains and strong intermolecular interaction between PEO chains through the hydrogen bonding [10]. The gradual decrease in intensity and increase of broadness of the diffraction peaks as a function of complexation of $NaIO_4$ salt(figure 1 (b) – (d)), reveal the decline in crystalline nature of blend electrolytes. This could be due to strong interaction of dissolved $NaIO_4$ salt with host polymer matrix and intermolecular interaction among the polymer chains of blend electrolytes, which lead to increase of amorphous phase in polymer blend electrolytes. It is note that (as evidenced from figure 1e) the intensity of characteristic XRD peaks corresponding to PEO further decreased by doping of 0.4 wt% of GO to 'PEO/PVP/$NaIO_4$(10 wt%)'electrolyte system. This evidences for further increase of amorphous portion in salt complexed electrolytes by the addition of GO, which favors for enhancement of mobility of the mobile charge ions in the polymer electrolytes and leads to an increase in the ionic conductivity [11].

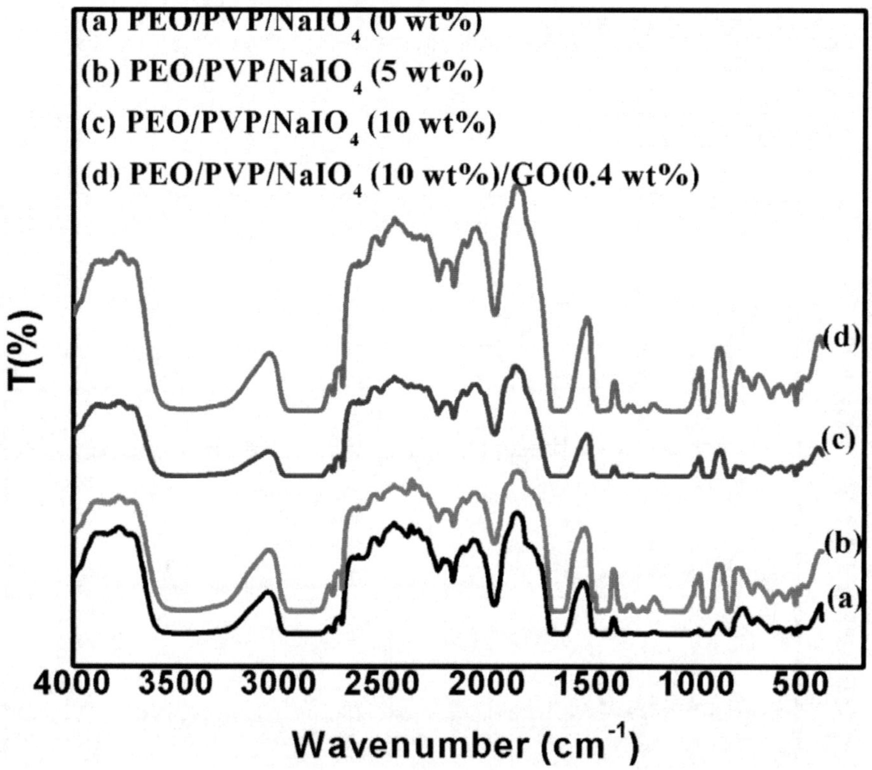

Figure 2. FTIR spectra for pure, $NaIO_4$ complexed and GO doped PEO/PVP blend electrolytes.

FTIR measurements are carried out for salt complexed and GO doped polymer blend electrolytes (as shown in figure 2) to understand their complexation properties such as ion–polymer interactions, intermolecular interactions, and their miscibility. The lower wavenumber region of800 cm^{-1} – 1000 cm^{-1} provides information about modifications in local structure of host polymer due to the interaction of the salt with polymer backbone. The vibrational modes present at 845 cm^{-1} and 947 $cm^{-1}$are accountable for CH_2 rocking motion with a little C–O stretching motion and C–O stretching motion with some contribution from CH_2 rocking motions, respectively (figure 2(a)). The intensities of two characteristic bands at 845 cm^{-1} and 947 cm^{-1} are associated to helical structure of PEO [12]. Considerable decrease in intensity of bands at 845 cm^{-1} and 947 cm^{-1} evidences for distortion in helical configuration of backbone of PEO polymer as a function of salt complexation. It gives the information of decrease in crystallinity and miscibility of ions with matrix of PEO/PVP blend. The small band present at 1350 cm^{-1} signifies the amorphous portion in PEO matrix, whose intensity increased with the increase of salt concentration. It evidences for increase of amorphous portion in the matrix of PEO polymer. In the FTIR spectra, the presence of all allowed vibrational modes and co-existence of well resolved respective band positions corresponding to ether oxygen groups (C–O–C) of PEO and carbonyl groups (C=O) of PVP are in congruent with previous reports in the literature and confirm the presence of PEO, PVP and their miscibility in resultant blend electrolyte films [13-15]. FTIR spectrum for GO doped 'PEO/PVP/$NaIO_4$ (10 wt%)' blend electrolyte film (figure 2(d)) displays a broad absorption band at around 3300 cm^{-1} due to typical carbonyl or carboxyl groups present in the GO structure [16]. As evidenced SEM studies, the surface of pure blend electrolyte (figure 3(a)) is observed to be relatively rough and composed with micro-void and cracks, which signifies the crystalline nature. The formation of micro-voids over surfaces films is due to rapid evaporation of the solvent during the preparation of polymer blends.

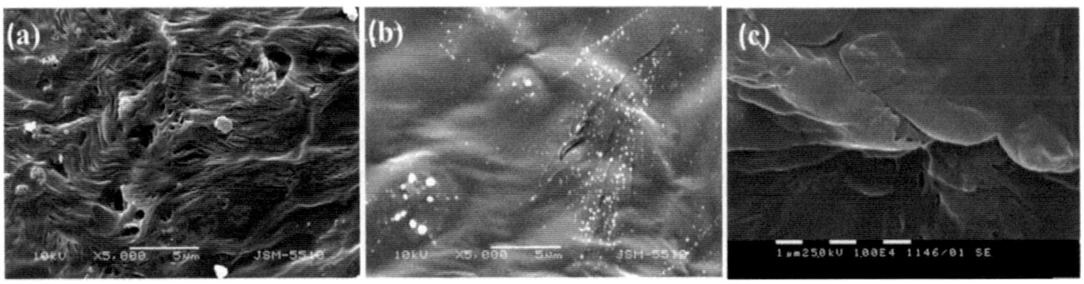

Figure 3. SEM images for $NaIO_4$ salt complexed and GO doped PEO/PVP electrolytes: (a) PEO/PVP blend, (b) PEO/PVP/$NaIO_4$ (10 wt%), (c) PEO/PVP/$NaIO_4$ (10 wt%)/GO (0.4 wt %)

Apparently, surfaces of the blend electrolyte films become smooth upon increasing of complexing salt concentration as shown in figure 3b. This indicates the increase in percentage of amorphous portion in the matrix of blend electrolytes due to random distribution and dissociation of salt. Significantly, the GO doped salt complexed polymer electrolytes demonstrated relatively smoother surface topology (figure 3(d)), it can be attributed to enhanced portion of amorphous region in the matrix of resultant salt complexed blend electrolyte due to addition of Graphene oxide.

3.2. Conductivity studies

Figure 4 shows Nyquist plots at room temperature for PEO/PVP polymer blend electrolytes complexed with $NaIO_4$ salt at various concentrations. In the plots, presence of depressed semicircles indicates non Debye nature of the polymer electrolytes and each semicircle represents the bulk resistance (R_b) of the respective electrolyte [17]. Nyquist plots are fitted with semicircle and the respective bulk resistance (R_b) of each polymer electrolyte is estimated from the point of intersection of higher end of semicircle with the real $(Z^{|} = R_b)$ axis [18]. The ionic conductivity (σ)of electrolyte films is estimated from following equation.

$$\sigma = \left({}^{t}/_{R_b} * A \right) \quad (1)$$

where t and A represent thickness and area of the polymer electrolyte, respectively. The estimated room temperature conductivity of pure PEO/PVP blend electrolyte is 2.24 X 10^{-9} S/cm and found to be increased to 1.57 X 10^{-7} S/cm as a result of increase in salt concentration to 10 wt%.

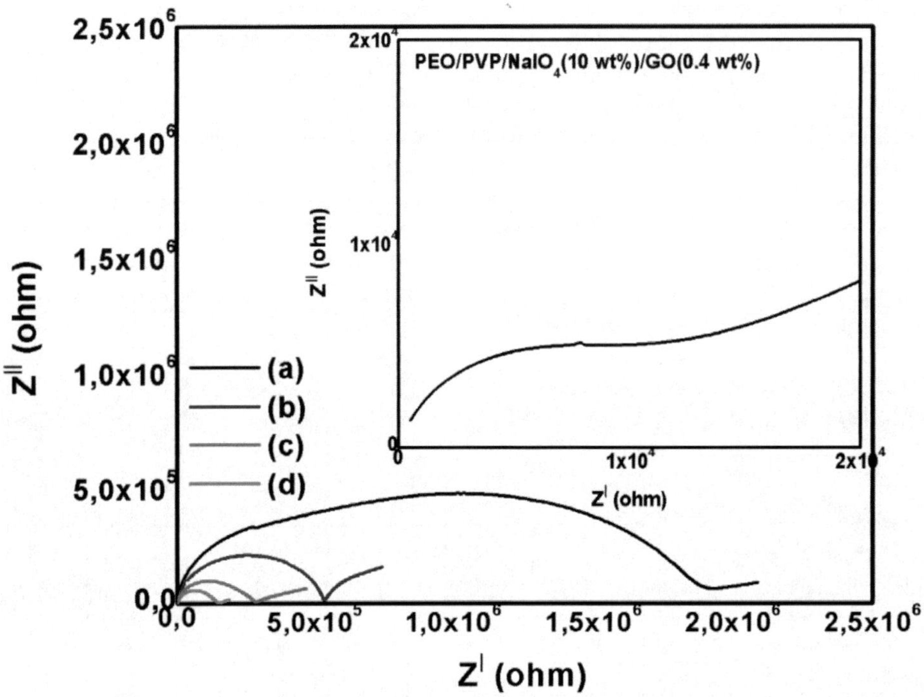

Figure 4. Nyquist plots at room temperature for $NaIO_4$ salt complexed PEO/PVP blend electrolytes: (a) PEO/PVP blend(b) PEO/PVP/$NaIO_4$ (5 wt%), (c) PEO/PVP/$NaIO_4$ (7.5wt%), (d) PEO/PVP/$NaIO_4$(10 wt%).

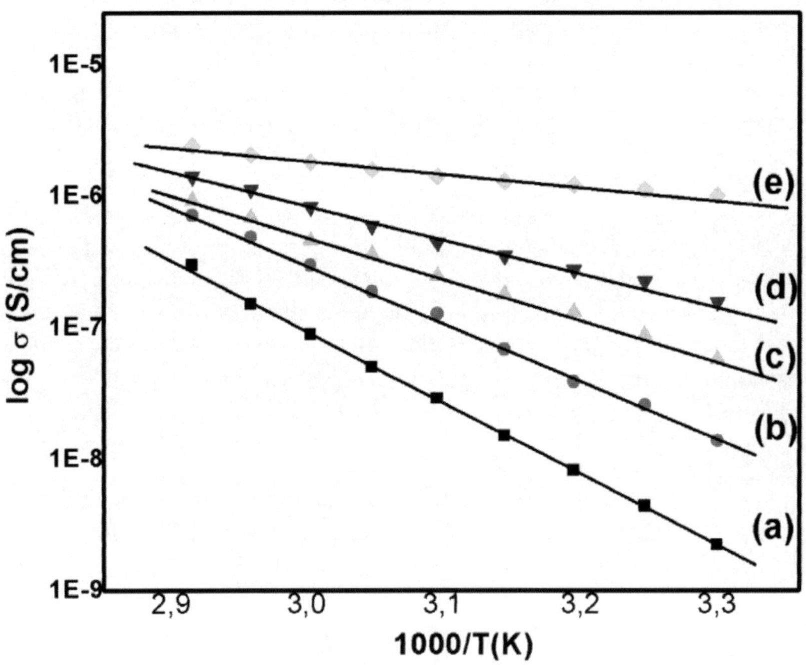

Figure 5. Arrhenius plots for $NaIO_4$ salt complexed and GO doped 'PEO/PVP blend' electrolytes: (a) PEO/PVP pure blend, (b)PEO/PVP/$NaIO_4$(5 wt%), (c) PEO/PVP/$NaIO_4$(7.5 wt%), (d) PEO/PVP/$NaIO_4$ (10wt%), (e) PEO/PVP/$NaIO_4$(10wt%)/GO(0.4 wt%).

This could be due to maximum PEO/PVP electrolyte uptake and enhanced coordination interactions of ether oxygen atoms of PEO or/and carbonyl oxygen atoms of PVP with Na^+ cations. This leads to increase amorphous content in PEO/PVP electrolyte and favour for increase of ionic conductivity [19].As a result of GO (0.4 wt%) doping, the intercept of higher end semicircle corresponding to 'PEO/PVP/$NaIO_4$ (10 wt%) electrolyte' on the $Z^|$ - axis tends move towards lower value as shown in inset of figure 4. The respective room temperature conductivity of 'PEO/PVP/$NaIO_4$(10 wt%)/GO(0.4 wt%)' electrolyte is found to be increase to 1 X 10^{-6} S/cm.This increase in Na^+-ion conductivity is due to the nature of graphene oxide that has a lot of oxygen-containing functionalgroups such as carboxylic acid, epoxy and hydroxyl, which can improve charge carrier mobilityand promote ionic conductivity. The large number of functional groups in GOenables it to dissolve more efficiently in methanol and distribute homogenously in PEO/PVP electrolytes. This could facilitate the mobility of mobile charge ions and leads to increase in ionic conductivity [20]. Figure 5 illustrates the variation in electrical conductivity of pure, salt complexed and GO doped polymer blend electrolytes as a function of temperature in the range 303–343 K. Evidently, the linear variation of resultant curves represents Arrhenius – type of thermally activated process by satisfying the following equation.

$$\sigma_{dc} = \sigma_0 exp(-E_a/KT), \qquad (2)$$

where σ_0 , E_a, K and T representthe pre-exponential factor, activation energy, Boltzmann constant and absolute temperatures, respectively. As a function of temperature, ionic conductivities of complexed blend electrolytes are increased due to enhanced polymer segmental mobility.

4. Conclusions

PEO/PVP blend electrolyte films complexed with $NaIO_4$ salt and doped with Graphene Oxide (GO), have been prepared by using conventional solution cast technique. The presence of characteristic X-ray diffraction peaks at 19.3^0 and 23.4^0 indicates semicrystalline nature of the pure blend electrolytes. Upon increase in salt concentration and inclusion of GO, decrease in intensity of diffraction peaks indicates increase of amorphous portion in the blend electrolytes due to strong interaction of dissolved additives with matrix of polymer electrolytes. In FTIR spectra, the coexistence of functional groups & vibrational bands corresponding to PEO and PVP reveals their uniform miscibility in blend electrolyte film and variation in the environment of the functional groups due to completion of salt and addition of GO. Surface smoothness of the blend electrolytes increased as a result of inclusion additives, which suggests the improvement in amorphous region in the matrix of polymer blends. The room ionic conductivity blend electrolytes increased with increase in salt concentration and found to be maximum of 1.57 X 10^{-7} S/cm for PEO/PVP/$NaIO_4$ (10 wt%). Upon inclusion of 0.4 wt% GO, resultant room temperature conductivity further increased to 1 X 10^{-6} S/cm, which signifies the improvement in flexibility of blend electrolytes due to interaction of GO with matrix of blend. Temperature dependent conductivity studies reveal increase in segmental motion, which supports for the enhancement of ionic conductivity in amorphous rich electrolytes.

Acknowledgements

The authors acknowledge the INERA (EU project Research Potential (FP7-316309-REGPOT-2012-2013-1) support for their participation at INERA Conference 2016 "Vapor Phase Technologies for Metal Oxide and Carbon Nanostructures", 5-9 July, 2016, Velingrad, Bulgaria.

References

[1] Jimin S, Dong-Gyun K, Hee J K, Jin H L, Ji-Hoon B and Jong-Chan L 2014 *J. Mater. Chem. A* **2** 13873
[2] Vignarooban K, Kushagra R, Elango A, Badami P, Mellander B.-E, Xu X, Tucker T G, Nam C, Kannan A M 2016 *Internaltional Journal of Hydrogen Energy* **41** 2829

[3] Pan H, Hu YS, Chen L 2013 *Energy Environ Sci* **6** 2338
[4] Armand M, 1994 *Solid State Ionics* **69** 309
[5] Fenton D F, Parker J M and Wright P V 1973 *Polymer* **14** 589
[6] Kiran K K, Ravi M, Pavani Y, Bhavani S, Sharma A K, Narasimha Rao V V R 2011 *Phys. B: Condens. Matter* **406** 1706
[7] Zardalidis G, Ioannou E, Pispas S and Floudas G 2013 *Macromolecules* **46** 2705
[8] Kiran K K, Ravi M, Pavani Y, Bhavani S, Sharma A K, Narasimha Rao V V R 2014 *Journal of Membrane Science* **454** 200
[9] Choi Y R, Yoon Y-G, Choi K S, Kang J H, Shim Y S, Kim Y.H, Chang H J, Lee J H, Park C R, Kim S Y 2016 *Carbon* **91** 178
[10] Agrawal R C, Hashmi S A 2007 *Ionics* **13** 295
[11] Yang X, Zhang L, Zhang F, Zhang T, Huang Y, Chen Y 2014 *Carbon* **72** 381
[12] WrightP V 1998 *Electrochimica Acta* **43** 1137
[13] Zain N M, Arof A K 1998 *Material Science Engineering B* **52** 40
[14] Simon G P, Shen Z, Cheng Y B 2003 *European Polymer Journal* **39** 1917
[15] Ramya C S, Selvasekranpandian S, Hiran Kumar G, Savitha T, Angelo P C 2008 *Journal of Non-Crystalline Solids* **354** 1494
[16] Cao Y C, Xu C, Wu X, Wang X, Xing L, Scott K 2011 *J. Power Sources* **196** 8377
[17] Lanfredi S, Saia P S, Lebullenger R 2002 *Solid State Ionics* **146** 329
[18] Mishra R, Baskaran N, Ramakrishnan P A, Rao K J 1998 *Solid State Ionics* **112** 261
[19] Park C H, Kim D W, Prakash J, Sun Y K 2003 *Solid State Ionics* **159** 111.
[20] Ye Y.-S, Rick J, Hwang B J 2013 *J. Mater. Chem. A* **1** 2719

Journal of Physics: Conference Series **764** (2016) 012007 doi:10.1088/1742-6596/764/1/012007

Antioxidant activity of membrane-fractionated coffee extracts in dependence of the storage conditions

D Mitev[1], D Peshev[2], G Peev[2] and L Peeva[3]

[1] REGPOT-2012-2013-1 NMP INERA Project, Institute of Solid State Physics-BAS, 72 Tzarigradsko Chaussee Blvd., 1784 Sofia, BULGARIA
[2] University of Chemical Technology and Metallurgy, Department of Chemical Engineering, 8 Kl. Ohridsky Blvd., 1756 Sofia, Bulgaria
[3] Chemical Engineering and Chemical Technology, Imperial College London, Exhibition Road, London SW7 2AZ, U.K.

Email: dpmitev@issp.bas.bg

Abstract. Present paper aims at one of the important aspects of the application of products with antioxidant activity: namely the preservation and change of their properties during the storage in different conditions, as well as their reliable characterisation.
The tests of antioxidant properties were conducted with membrane-separated coffee extracts, isolated using a "Microdyn Nadir NP030P" type of commercial nanofiltration membrane (30% retention of NaCl; MWCO~400). Prepared coffee permeates and retentates were stored 0÷10 days in cool/warm conditions, with/without air access and at different illumination conditions. The kinetics of content changes was evaluated according to Folin-Ciocalteu method of total phenolic/reducing content determination.

1. Introduction

The effects of storage conditions on the stability of different herbal, biological, medical, nutritional etc. extracts are widely researched; with objects of interest being: from plants and mushrooms, foods, flavourings, blood and tissues, DNA extracts, to house dust mites and many more [1-5]. Nowadays a lot of attention is attracted to the membrane-derived plant products, especially in aspect of their usage as alternative natural remedies and healthy food supplements [6-7]. As regards the application of products with antioxidant activity, here the preservation and as minor as possible change of their properties during the storage in different conditions plays a very important role. Any research of biologically-active extracts, solutions, samples etc. and their deterioration products relies on a reliable and possibly simple characterisation.

With present work we exhibit a practical application of the nanomembrane filtration, aimed at the important task of durability assessment for the herbal extracts and their nanofiltration derivatives. Analytical part is comprised by a determination of the antioxidant efficiency of coffee extracts under storage, implementing the Folin-Ciocalteu's oxidative method and spectrophotometric evaluations.

2. Experimental part

2.1. Materials

All reagents used were of analytical grade quality: Folin & Ciocalteu's phenol reagent, gallic acid, anhydrous sodium carbonate from Sigma-Aldrich, Taufkirchen, Germany. Syringe filters of 0.45 μm

(Valerus Ltd., Sofia, Bulgaria). Distilled water (water still GFL Typ 2004, Burgwedel, Germany) was used throughout the work.

2.2. Coffee extract preparation

A short brew of coffee type "Carraro" (Belogia Espresso Machine) was used for the preparation of the initial extract. An attempt to eliminate hampering the subsequent membrane filtration colloidal coffee ground particles through centrifugation at 3000 min^{-1} for 20 minutes turned out to be insufficient. Therefore as a primary treatment filtration using "Boeco" 391 filter paper (2-3 µm porosity) was applied, followed by 0.45 µm syringe filters. Filtered this way extract was finally diluted twentyfold and directed to the nanomembrane separation.

2.3. Nanomembrane filration

150 mL of the diluted coffee extract (further: Feed; F) were subjected to a nanofiltration (set-up according to 2.6.; filtration procedure as described previously [6]). The membrane separation continued until 76 mL of permeate (P) were isolated, with 74 mL of retentate (R) being the rest. The nanomembrane performance is shown in the subsection 3.1.

2.4. Preparation of samples for "durability kinetics" tests

Samples were divided in four major groups and two sets (figure 1): 1) storage with presence of air in cool conditions (samples F_{1S}, P_{1S} and R_{1S}); 2) storage without air presence in cool conditions (samples F_{1Sn}, P_{1Sn} and R_{1Sn}); 3) storage with presence of air in warm illuminated conditions (samples F_{1T}, P_{1T} and R_{1T}); 4) storage without air presence in warm illuminated conditions (samples F_{1Tn}, P_{1Tn} and R_{1Tn}). Before mentioned samples represented also the first set, which was incubated until the sixth day from the deposition; next set was researched on the tenth day (samples F_{2S}, P_{2S} and R_{2S}; F_{2Sn}, P_{2Sn} and R_{2Sn}; F_{2T}, P_{2T} and R_{2T}; F_{2Tn}, P_{2Tn} and R_{2Tn}). The samples for no air presence were packed in 2-mL HPLC vials; those for air presence had 5-mL samples poured in 50-mL plugged volumetric flasks. The storage for cool conditions was executed at 8°C in refrigerator in dark; the storage for warm conditions was at room temperature around 20°C, near a window open to light as shown on the figure 1.

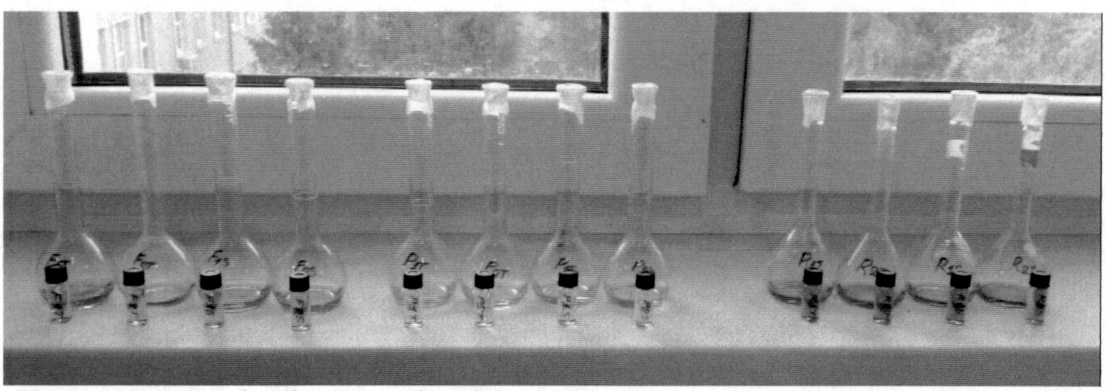

Figure 1. Samples of coffee extract (from left): feed, permeate and retentate prepared for durability tests

2.5. Total phenolic content (TPC) determination

A spectrophotometric assay of TPC was applied using a modified method of Folin-Ciocalteu method [8-9]. The assay in sum was following: corresponding amount of the phenolic/reductive coffee extract (0.200 to 1.000) was brought to a volume of 7 mL through appropriate dilution with distilled water. 0.5 mL of of Folin–Ciocalteu's reagent was subsequently added, and after thorough shaking 10 minutes of dead time was applied. Then 2.5 mL of 10% (w/w) aqueous sodium carbonate solution was added. The colourisation reaction developed for 20 min at room temperature. Afterwards the specific absorbance at 760 nm was finally measured at room temperature with a UV–Vis spectrophotometer;

an identical mixture without coffee extract was used as a blank sample. The total phenolic/reducing contents were rated as mg kg^{-1} gallic acid equivalents (GAE), through a calibration curve of gallic acid (figure 2) and normalised per kg of initial extract sample.

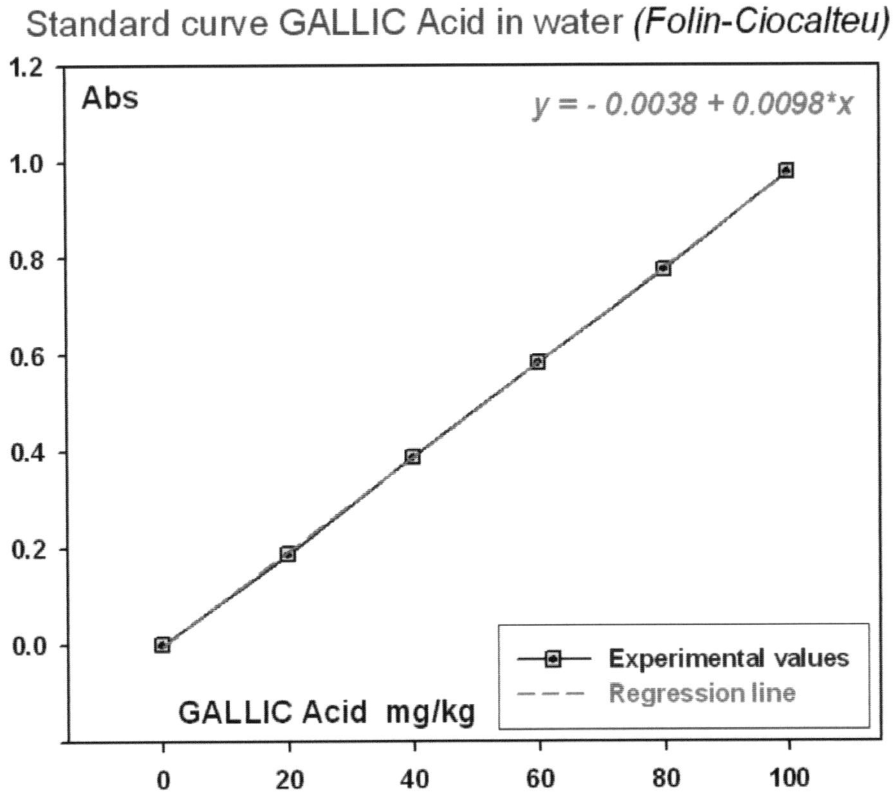

Figure 2. The standard curve for the gallic acid standard. Rsqr = 0.9999

2.6. Instruments

Following instrumentation was used throughout the work:

Membrane filtration. "Dead-end" filtration cell of type "METcell" (Membrane Extraction Technology, London, UK), fitted with nanofiltration membrane "Microdyn Nadir" NP 030 P (Wiesbaden, Germany); processing at 10 bar (compressed nitrogen).

Spectrophotometry. An S-22 UV/Vis type of spectrophotometer (Boeco, Germany) was used for determination of the maximal spectral absorption at wavelength of 760 nm.

Centrifugation. "Janetzki" T32a centrifuge (Berlin, Germany).

3. Results and Discussion

Extracts having different antioxidant properties and degree of degradation were researched with the course of our work. In addition to the different duration and conditions of each procedure, the samples' response was specific also for each type of used feed or nanofiltration product.

3.1. Nanomembrane filtration of initial coffee extract

Membranes used in the "production scale" filtration procedures are exclusively operated under crossflow filtration mode [10]. For the laboratory practice though membranes are mostly operated under "dead-end" filtration mode, simpler and faster approach with no parallel feed flows onto the membrane surface. Under the dead-end filtration mode, all the colloidal particles and high-molecular compounds contained in the feed coffee extract deposit on the membrane surface, in addition to the potential concentration polarisation. Both phenomena result in permeate flow rate reduction with working solutions, compared to filtration runs with pure solvent. This effect is illustrated in figure 3 for the used nanomembrane type "Microdyn Nadir" NP 030 P, where a cubic regression was the best fit for coffee extract nanofiltration kinetics. The membrane rejection in regard to the antioxidant contents of the feed extract was calculated in two different ways [6], with rejected compounds mainly over 350-400 Da molecular weight:

$$R_1 = (1 - C_p/C_f)*100 \quad = 77.2\%, \text{ or} \tag{1}$$

$$R_2 = (lg(C_r/C_f)/lg(V_f/V_r))*100 \quad = 75.8\% \tag{2}$$

The membrane performed within the manufacturers specifications for water permeate flux: 2.4913 LMH/bar (L/(m^{2}hbar)) before the coffee extract filtration and 1.8519 LMH/bar after it; with manufacturers' data being >1 LMH/bar.

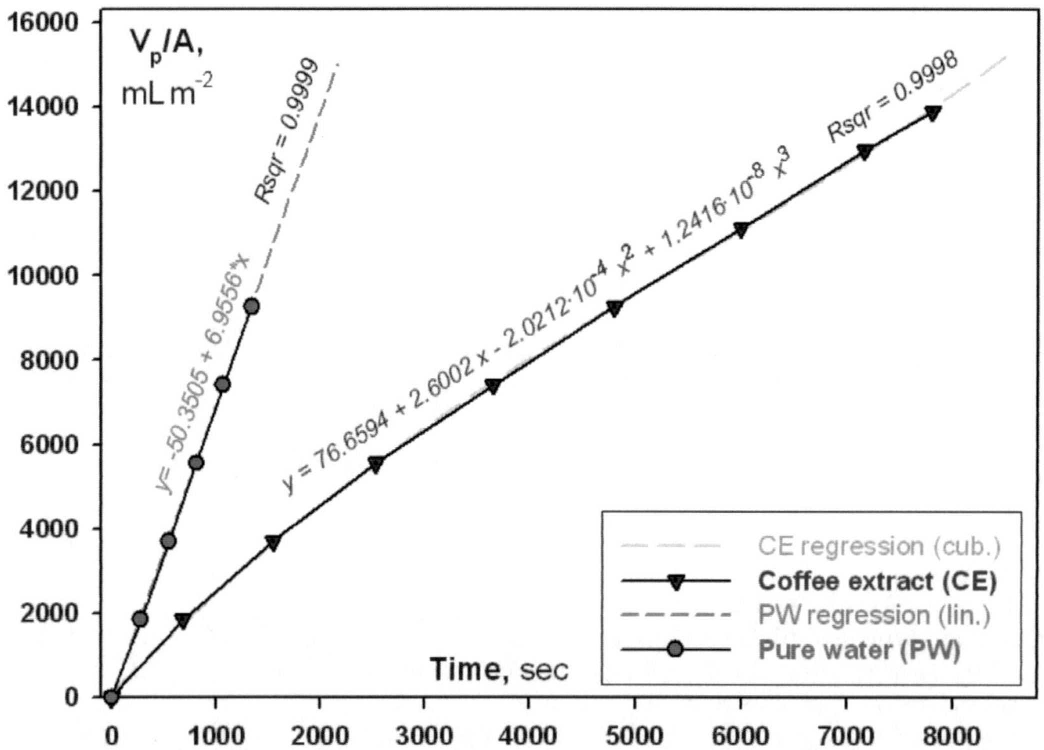

Figure 3. Kinetics of the NP 030 P nanofiltration of pure water and coffee extract: cumulative permeate volume per unit membrane surface versus the time of collection

3.2. The "Durability Kinetics"

According to the Folin-Ciocalteu tests, the primary short "Carraro" coffee brew contained 3722 mg kg^{-1} GAE total polyphenolic and reducing content of antioxidants. Diluted feed extract reached C_f=182.4 mg kg^{-1} GAE, with permeate and retentate immediately after nanofiltration C_p=41.6 and C_r=308.5 mg kg^{-1} GAE. After six days of incubation period relatively limited changes with the antioxidant capacity were observed; with bigger downshift for both of permeates and especially for the P_{1T} which was kept in warm illuminated conditions. Excluding the latter, less than 2% RSD dispersion was registered for this first set (Table 1 and figure 4).

Table 1. Kinetics of the antioxidant contents in feed coffee extract, permeate and retentate at sixth and tenth day of storage; RSD here depicts the degree of changes

Sample type	AIR	No AIR	RSD,%	**Sample type**	AIR	No AIR	RSD,%
F_{1T}	173.5	175.9	0.97	F_{2T}	165.4	169.9	1.90
P_{1T}	33.4	39.7	12.19	P_{2T}	27.9	38.3	22.22
R_{1T}	286.8	293.8	1.71	R_{2T}	191.3	289.3	28.84
F_{1S}	183.7	185.6	0.73	F_{2S}	173.9	182.9	3.57
P_{1S}	38.8	39.9	1.98	P_{2S}	36.4	39.4	5.60
R_{1S}	301.0	301.9	0.21	R_{2S}	282.2	297.8	3.80

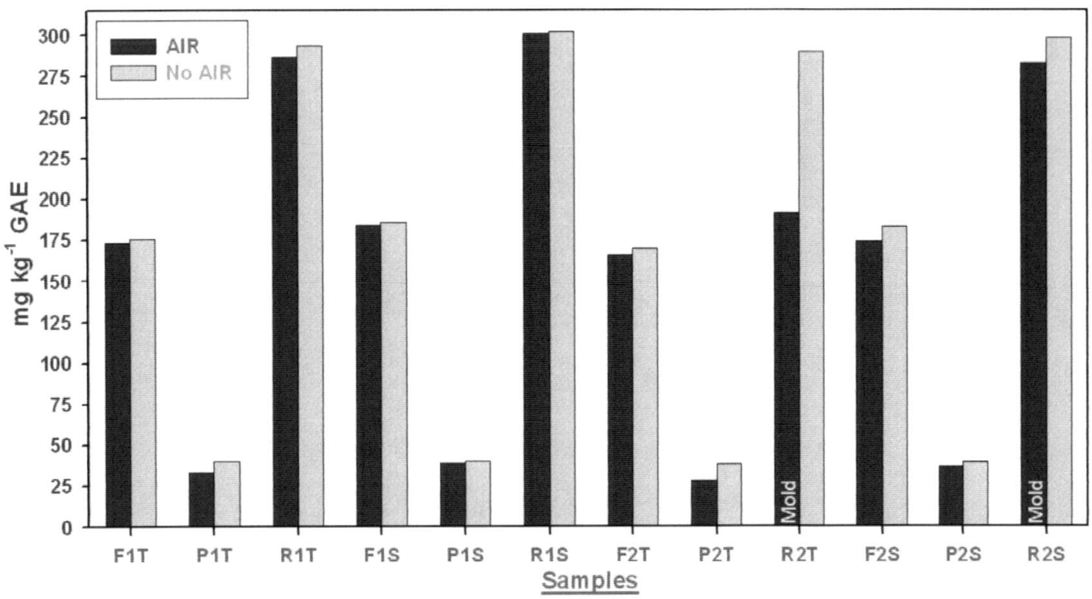

Figure 4. Comparative kinetics for the two sets of samples.

Significant changes were logged at the tenth day of incubation period though. Exposed and kept in warm samples showed rapid deterioration, except the feed extract. A possible reason for the feed extraxts' stability is the preliminary sterilization during the brewing process. For both of retentates some mold cultures started to develop into the air-containing flasks, more abundant for R_{2T}, where TPC value plummeted with a third. Despite of the mold presence, the TPC for the deposited in cold R_{2S} was not significantly affected. The growth of the mold was most likely promoted by the retained carbohydrates in R-samples with the nanofiltration; but this must be further specified.

4. Conclusions

With this work we studied the nanofiltration of the coffee brew extract, and changes of the derived products under different duration and storage conditions. Membrane filtration here exhibited a strong necessity of preliminary micro- and ultra-filtration steps, in order nanomembrane fouling to be avoided.

All in all, the extract samples showed small antioxidative changes until the sixth day of storage irrespective of the conditions; further storage could be considered reasonable only at lower temperatures and especially without air in presence.

Acknowledgements
REGPOT-2012-2013-1 NMP INERA Project is kindly acknowledged. The authors also acknowledge University of Chemical Technology and Metallurgy – Sofia for financial support of this research (Project No. 11572).

References

[1] Ndungutse V, Mereddy R and Sultanbawa Y 2015 Bioactive Properties of Mushroom (Agaricus bisporus) Stipe Extracts *Journal of Food Processing and Preservation* **39** (6) 2225-2233
[2] Ranveer RC, Gatade AA, Kamble HA and Sahoo AK 2015 Microencapsulation and Storage Stability of Lycopene Extracted from Tomato Processing Waste *Brazilian Archives of Biology and Technology* **58** (6) 953-960
[3] Charde MS, Ingole P, Welankiwar AS, Kumar J and Chakole RD 2014 Review: The procurement, storage and quality assurance of frozen blood and tissue biospecimens *International Journal of Pharmacological Research* **4** (2) 49-52
[4] Lee SB, Crouse CA and Kline MC 2010 Optimizing Storage and Handling of DNA Extracts *Forensic Sci Rev.* **22** (2) 131-44
[5] Jeong KY, Choi S-Y, Han I-S, et al. 2013 The Effects of Storage Conditions on the Stability of House Dust Mite Extracts *Allergy, Asthma & Immunology Research.* **5** (6) 397-401
[6] Peev G, Penchev P, Peshev D, Angelov G 2011 Solvent extraction of rosmarinic acid from lemon balm and concentration of extracts by nanofiltration: Effect of plant pre-treatment by supercritical carbon dioxide *Chemical Engineering Research & Design* **89** 2236-2243
[7] Tylkowski B, Trusheva B, Bankova V, Giamberini M, Peev G, Nikolova A 2010 Extraction of biologically active compounds from propolis and concentration of extract by nanofiltration *Journal of Membrane Science* **348** (1-2) 124-130
[8] Singleton VL, Orthofer R and Lamuela-Raventós RM 1999 Analysis of total phenols and other oxidation substrates and antioxidants by means of Folin–Ciocalteu reagent *Methods in Enzymology* **299** 152-178
[9] Singleton VL and Rossi JA, 1965 Colourimetry of total phenolics with phosphomolybdic–phosphotungstic acid reagents *American Journal of Enology and Viticulture* **16** 144-158
[10] Yoon S-H 2016 *Membrane Bioreactor Processes: Principles and Applications* (CRC Press, Boca Raton, FL, USA) 1-31

Journal of Physics: Conference Series **764** (2016) 012008 doi:10.1088/1742-6596/764/1/012008

Two-phonon Raman scattering in graphene for laser excitation beyond the π-plasmon energy

Valentin N Popov

Faculty of Physics, University of Sofia, BG-1164 Sofia, Bulgaria

E-mail: vpopov@phys.uni-sofia.bg

Abstract. The study of the two-phonon Raman spectra of graphene allows obtaining valuable information about the phonon dispersion inside the Brillouin zone. Up to now, the two-phonon Raman spectra have been extensively measured and modelled for laser excitation mainly in the visible range. However, recently, Raman measurements have been reported for laser excitation in the near ultraviolet range, close to the π-plasmon energy of about 5 eV. Here, we calculate the two-phonon Raman spectra in the latter range and compare the obtained results to the available Raman data. The crossover behaviour of the scattering of electrons by phonons on passing the π-plasmon energy is also discussed.

1. Introduction

Graphene has a two-dimensional honeycomb structure entirely consisting of carbon atoms [1] (figure 1). Its unique mechanical, optical, and electronic properties [2,3] make it suitable for application for thin transparent conducting electrodes, and flexible electronics and photonics [4]. Among the various experimental techniques, which are currently applied for the characterization of graphene samples, the Raman spectroscopy has become most popular because of its obvious advantages as an inexpensive and non-destructive method [5].

Graphene has a single Raman-active phonon, which gives rise to the so-called G band in the Raman spectra. A few peaks are normally observed in the second-order Raman spectra of graphene. In perfect graphene, the spectra are dominated by the so-called 2D band, which is several times more intense than the G band. It arises from scattering of electrons by phonons close to the K point in the Brillouin zone (figure 2). A few other two-phonon bands are also usually seen in the spectra. In the presence of defects, defect-induced Raman bands, which come from scattering by a defect and a phonon in the vicinity of the K point, are often detected. The two-phonon bands contain valuable information about the phonon dispersion, the electron-phonon and electron-photon couplings, doping, and strain in perfect graphene [6]. The G band and the second-order Raman bands originate from *resonant Raman scattering*, which has been attributed to the unique electronic band structure of graphene, consisting of bands with linear dispersion at the Fermi energy [7].

While the modelling of defect-induced bands relies on the unknown defect type and distribution, and significant advance in their quantitative description has not been achieved so far, there have been a number of reports on progress in the computation of the two-phonon bands for laser excitation in the visible range, predicting realistic position, lineshape, and dispersion rate of these bands [8,9].

Journal of Physics: Conference Series **764** (2016) 012008 doi:10.1088/1742-6596/764/1/012008

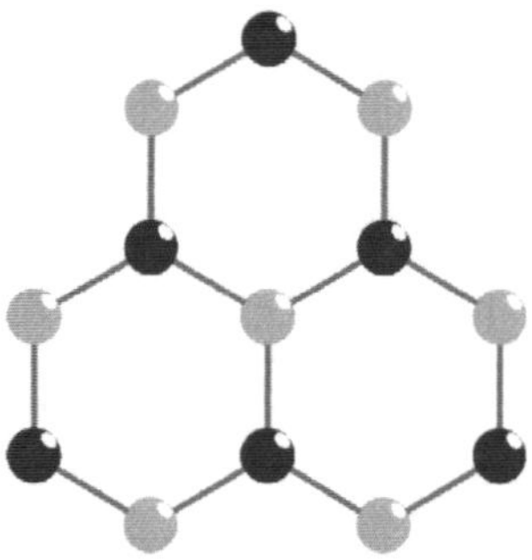

Figure 1. The atomic structure of graphene. The atoms of the two interpenetrating hexagonal lattices are drawn by light and dark grey circles.

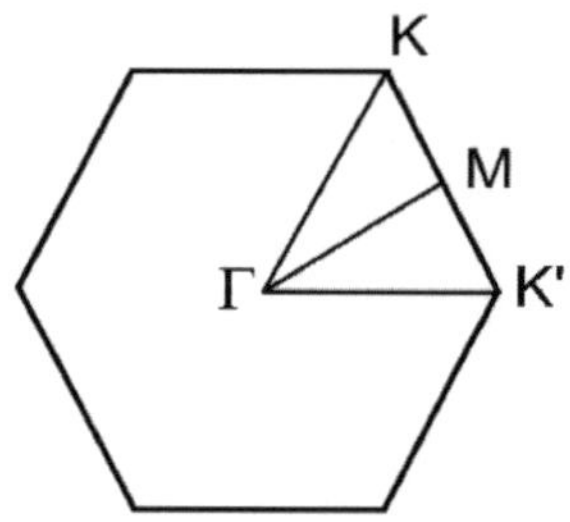

Figure 2. The hexagonal Brillouin zone of graphene with indicated special points.

Here, we extend the calculations of the two-phonon bands of Ref. [9] to laser excitation above the optical range and even beyond the π-plasmon energy of graphene, prompted by recent measurements of the Raman bands in this energy range [10]. We limit our report to two-phonon bands of in-plane optical phonons of graphene, which yield most intense Raman peaks. The paper is organized as follows. First, the underlying theory of the electron and phonon dispersions, as well as the resonant Raman intensity, is briefly dwelt on in Sec. 2. The obtained results for the overtone and combination bands are presented and discussed in Sec. 3. The paper ends up with conclusions.

2. Theoretical part

The resonant enhancement of the two-phonon Raman bands is a result of matching the incident photon energy with quantum transitions of the studied system. Therefore, the consistent modelling of the Raman bands requires quantum-mechanical description of the system of electrons, photons, phonons, and their interactions. Here, the calculation of the electronic band structure, phonon dispersion, and Raman intensity is performed within a density-functional-based non-orthogonal tight-binding model (NTB model) [11]. The NTB model uses Hamiltonian and overlap matrix elements with radial dependence derived from an *ab-initio* study on carbon dimers [12]. Only matrix elements between the four atomic orbitals of type s, p_x, p_y, p_z for the valence electrons of the carbon atoms are considered. The angular dependence of the matrix elements is described within the Slater – Koster scheme. The NTB model allows for the calculation of the total energy of the atomic structure with important application to the relaxation of the atomic structure, which is a necessary step before performing calculations of the phonon dispersion.

The phonon dispersion is derived using a perturbative approach within the NTB model [13]. The expression for the dynamical matrix makes use of the electron-phonon matrix elements, which are also derived within the NTB model [14].

Finally, the Raman intensity of a two-phonon Raman band can be calculated by means of the expression, obtained in fourth-order perturbation theory within the NTB model [9],

$$I \propto \sum_f \left| \sum_{c,b,a} \frac{M_{fc} M_{cb} M_{ba} M_{ai}}{\Delta E_{ic} \Delta E_{ib} \Delta E_{ia}} \right|^2 \delta\left(E_i - E_f\right). \quad (1)$$

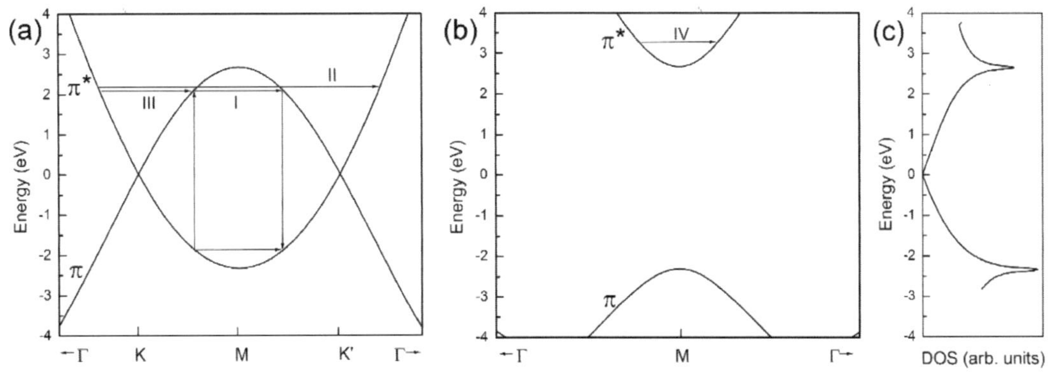

Figure 3. Electronic band structure of graphene along the high-symmetry directions: (a) ΓKMK′Γ and (b) ΓMΓ. Only the conduction π* band and valence π band close to the Fermi energy, taken to be zero, are shown. These bands have linear dispersion in the vicinity of the Fermi energy but have inflexion points at the M point. The electronic DOS, shown in panel (c), has logarithmic singularities at the inflexion points. For explanation of the notation, see text.

Here, $\Delta E_{iu} = E_i - E_u - i\gamma$, where E_u ($u = a,b,c$) are the energies of the intermediate states of the system of photons, electrons, holes, and phonons; E_i is the initial state energy, equal to the laser excitation energy E_L; γ is the sum of the halfwidths of pairs of electron and hole states, belonging to an intermediate system state, and will be referred to as the *electronic linewidth*; the approximation of wavevector-independent but energy-dependent electronic linewidth is adopted throughout the paper; M_{ai} and M_{fc} are the momentum matrix elements for creation and recombination of an electron-hole pair, respectively; M_{ba} and M_{cb} are the electron-phonon matrix elements for scattering between intermediate states. The electron-photon and electron-phonon matrix elements are calculated explicitly within the NTB model [14]. The summation over the intermediate states runs over all valence and conduction bands, and over all electron wavevectors **k**. The summation over the final states runs over the considered pair of phonon branches and over all phonon wavevectors **q**. For both summations, convergence is reached with a 400 × 400 mesh of **k** and **q** points in the Brillouin zone. The calculations are limited to in-plane parallel scattering configuration, backscattering geometry, and Stokes processes.

As it is seen from equation (1), the Raman intensity is resonantly enhanced, whenever in the sum there are terms, for which the real part of one, two, or three of the quantities ΔE_{iu} vanishes, i.e., the laser excitation matches well intermediate states of the system or, roughly speaking, matches the difference of the energies of the created electron and hole. Additionally, the Raman bands will be observable if the matrix elements in equation (1) are sufficiently large. On the contrary, terms, for which none of the energy denominators ΔE_{iu} is small, will give negligible contribution to the intensity, even if the matrix elements are not small.

3. Results and discussion

The honeycomb atomic structure of graphene has a unit cell of two atoms and can be considered as consisting of two interpenetrating hexagonal Bravais lattices (figure 1). Graphene has a hexagonal Brillouin zone with several characteristic special points (figure 2), among which the points Γ, M, K, and K′ are most important for this study as it will become clear below.

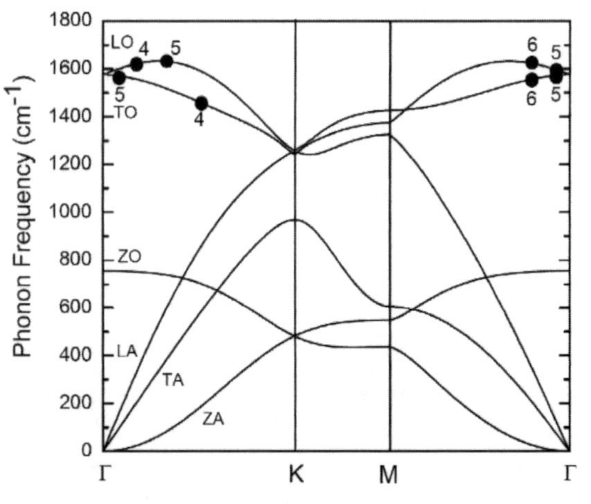

Figure 4. Phonon dispersion of graphene with displacement pattern for the branches. The circles mark the phonon wavevectors for two-phonon scattering at laser excitation, given by the numbers next to the symbols. For explanation of the branch notation, see text.

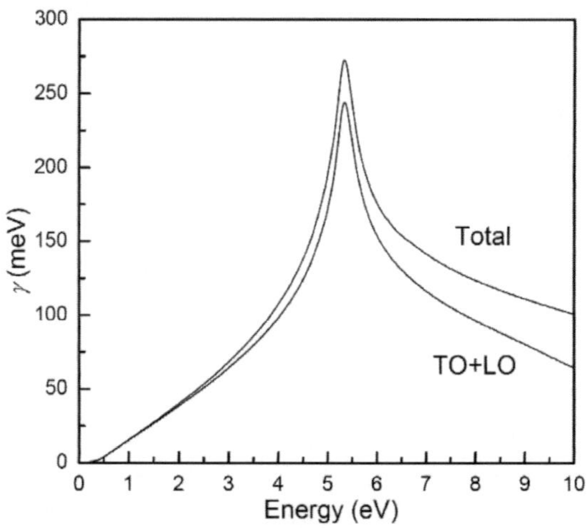

Figure 5. Total electronic linewidth and contribution to the linewidth from LO and TO phonons only. The predominant contribution of the latter phonons to the linewidth is due to the strong coupling of electrons with these phonons.

The electronic band structure, derived by solving the non-orthogonal tight-binding wave equation along the high-symmetry directions ΓKMK′Γ and ΓMΓ, is shown in figures 3(a) and 3(b). In the vicinity of the Fermi energy, chosen to be zero, only the conduction π* band and valence π band, arising from the p_z orbitals (the z axis is perpenducular to the plane of graphene) are present. An amazing feature of the band structure is the linear dispersion of the two bands close to the Fermi energy at the K and K′ points. The linear bands form cones above and below the zero-energy **k**-vector plane, so-called *Dirac cones*. In the effective-mass approximation, the effective-mass wave equation has the form of the wave equation of Dirac – Weil for massless fermions, which is often used in the simplified description of transport and optical phenomena in graphene.

Away from the Fermi energy, there is a deviation from the linear behaviour of the bands, known as the band *warping effect*. At energies ±2.5 eV, the bands are strongly modified between the K and K′ points and have inflexion points at the M point. The behaviour of the bands around the inflexion points is especially important for this study because of the crossover behaviour of the electronic scattering mechanism upon crossing these points and because of the enhancement of the Raman intensity in the vicinity of these points due to the logarithmic singularity of the electronic density of states (DOS) [15] (figure 3(c)).

The phonon dispersion is obtained from the dynamical equation for graphene (figure 4). The phonon branches are marked by symbols, characterizing their vibrational pattern. The longitudinal, transverse, and out-of-plane branches are denoted by the letters "L", "T", and "Z", respectively. A second letter "A" and "O" is added to the branch notation with the meaning of acoustic and optical, respectively. While all branches give contribution to the two-phonon bands, here we are particularly concerned with the LO and TO branches, from which the most intense Raman lines arise.

Finally, the energy dependence of the electronic linewidth is calculated by considering all electron-phonon scattering processes in the entire Brillouin zone (figure 5). As it is seen in figure 5, the LO and TO branches are most effective in scattering electrons, which is in accord with the result of the complete calculation (not provided here) that the remaining branches produce two-phonon Raman bands with relatively low intensity.

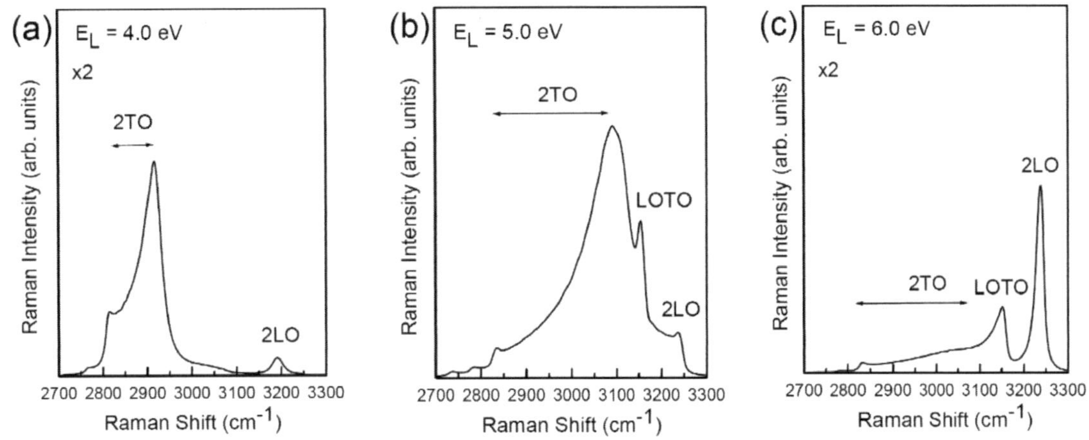

Figure 6. Calculated two-phonon Raman bands at: (a) $E_L = 4.0$ eV, (b) $E_L = 5.0$ eV, and (c) $E_L = 6.0$ eV.

Before proceeding with the results for the two-phonon bands, we note that the expression within the absolute value bars of equation (1) is the *scattering amplitude*. It describes *scattering processes*, consisting of virtual processes of electron-hole creation, scattering of an electron (or a hole) by a phonon, a consecutive scattering of an electron (or a hole) by a phonon with opposite momentum with regard to the first phonon, and final electron-hole annihilation. There are altogether eight different scattering processes and all of them are taken into account here. A simplified schematic of one of the eight scattering processes in graphene is displayed by a closed rectangle drawn by horizontal and vertical arrows in figure 3(a). The vertical arrows depict electron-hole creation or annihilation. For resonant processes, the vertical lines roughly correspond to the laser excitation. The horizontal arrows depict the phonon wavevectors for electron-phonon scattering processes I, II, III, and IV (figures 3(a) and 3(b)).

The calculated two-phonon Raman spectrum at $E_L = 4.0$ eV and in the frequency range between 2700 and 3300 cm^{-1} is shown in figure 6(a) and the participating phonons are marked by circles in figure 4. The intense band between 2800 and 2900 cm^{-1} comes from scattering by TO phonons close to the K point and is referred to as the 2TO band or the 2D band. The major contribution to this band (right peak) is due to processes I, while a minor contribution (left kink) is due to processes II. The asymmetry of this band can be attributed to the asymmetry of the TO branch with respect to the K point. The weak band at 3200 cm^{-1} arises from processes III triggered by LO phonons close to the Γ point and is known as the 2LO band.

With increasing E_L beyond the π-plasmon energy, process I disappears and a new process IV becomes effective. Recently, similar crossover behaviour has been predicted for the isostructural material silicene [16].

The two-phonon Raman spectrum at $E_L = 5.0$ eV, shown in figure 6(b), exhibits a broad Raman feature between 2800 and 3050 cm^{-1}, which is the 2TO band. The peak at 3050 cm^{-1} originates from processes I involving TO phonons close the Γ point and the kink at 2800 cm^{-1} originates from processes II with TO phonons close to the K point. The high-frequency kink at about 3250 cm^{-1} comes from processes III involving LO phonons close to the maximum ("overbending") of the LO branch (2LO band). At this laser excitation, a new scattering process IV appears (see figure 3(b)). The sharp line at 3150 cm^{-1} is due to processes IV with participation of LO and TO phonons close to the Γ point (a combination LOTO band). The overall shape of the Raman bands corresponds very well to the experimentally observed one [10], which supports the applicability of the applied computational approach to the calculation of the two-phonon bands of graphene.

The two-phonon Raman spectrum at $E_L = 6.0$ eV shows a broad band and two sharp bands arising from processes IV with LO and TO phonons in the vicinity of the Γ point. The former is the 2TO band, and the latter are the combination LOTO band and the 2LO band.

We note that the position of the 2TO, 2LO, and LOTO bands depends on the laser excitation, which is most strongly expressed in the case of the 2TO band. The reason for this dependence is that the condition for resonance interconnects the wavevectors of the electron and the phonon, as well as the difference of the energy of the electron and hole, the phonon energy, and the laser excitation. The derived average slope of this dependence (termed *dispersion rate*) for the 2TO band of 150 cm^{-1}/eV in the energy range 2.0 – 5.0 eV agrees well with the experimental one of 165 cm^{-1}/eV [10].

Finally, the enhancement of the Raman intensity close to the π-plasmon energy of about 5 eV (figure 6(b)) can be attributed to the singularities of the electronic DOS at the M point (see also Ref. [16]).

4. Conclusions

We have shown that the position of the two-phonon Raman bands can be associated with phonons inside the Brillouin zone. These predictions can be used for reconstruction of the phonon dispersion of graphene on the basis of experimental two-phonon Raman data.

Acknowledgments

The author would like to acknowledge the financial support from EU Seventh Framework Programme project INERA under grant agreement number 316309.

References

[1] Geim A K and Novoselov K S 2007 *Nature Materials* **6** 183
[2] Castro Neto A H, Guinea F, Peres N M R, Novoselov K S and Geim A K 2009 *Rev. Mod. Phys.* **81** 109
[3] McCann E and Koshino M 2013 *Rep. Prog. Phys.* **76** 056503
[4] Bonaccorso F, Sun Z, Hasan T and Ferrari A C 2010 *Nature Photonics* **4** 611
[5] Malard L M, Pimenta M A, Dresselhaus G and Dresselhaus M S 2009 *Phys. Rep.* **473** 51
[6] Ferrari A C and Basko D M 2013 *Nature Nanotechnology* **8** 235
[7] Thomsen C and Reich S 2000 *Phys. Rev. Lett.* 85 5214
[8] Venezuela P, Lazzeri M and Mauri F 2011 *Phys. Rev.* B **84** 035433
[9] Popov V N and Lambin Ph 2012 *Eur. Phys. J.* B **85** 418
[10] Tyborski C, Herziger F, Gillen R and Maultzsch J 2015 *Phys. Rev.* B **92** 041401(R)
[11] Popov V N and Henrard L 2004 *Phys. Rev.* B **70** 115407
[12] Porezag D, Frauenheim T, Köhler T, Seifert G and Kaschner R 1995 *Phys. Rev.* B **51** 12947
[13] Popov V N and Lambin Ph 2006 *Phys. Rev.* B **73** 085407
[14] Popov V N, Henrard L and Lambin Ph 2005 *Phys. Rev.* **72** 035436
[15] Hove L V 1953 *Phys. Rev.* **89** 1189
[16] Popov V N and Lambin Ph 2016 *2D Materials* **3** 025014

Journal of Physics: Conference Series **764** (2016) 012009 doi:10.1088/1742-6596/764/1/012009

Rejuvenation of degraded electrochromic MoO_3 thin films made by DC magnetron sputtering: Preliminary results

Miguel A Arvizu, Claes G Granqvist and Gunnar A Niklasson

Department of Engineering Sciences, The Ångström Laboratory, Uppsala University, P. O. Box 534, SE-751 21 Uppsala, Sweden

E-mail: Miguel.Arvizu@angstrom.uu.se

Abstract. Molybdenum oxide thin films were deposited by reactive DC magnetron sputtering and were subjected to voltammetric cycling in an electrolyte comprised of lithium perchlorate in propylene carbonate. The films were heavily degraded during 20 voltammetric cycles in an extended voltage range. The films were subsequently rejuvenated by use of potentiostatic treatments under different voltages during 20 hours. Optical changes were recorded during the electrochemical degradation and ensuing rejuvenation.

1. Introduction

Electrochromic (EC) materials are capable of going through reversible and persistent optical changes when subjected to a potential difference exerting insertion/extraction of ions and electrons [1]. Some transition metal oxides exhibit remarkable EC properties and are excellent candidates for use in technical applications such as "smart windows", *i.e.*, devices intended to allow a user to control the throughput of visible light and solar energy into buildings, thereby making these devices an excellent option for energy saving purposes [2].

Electrochromism in molybdenum oxide (MoO_3) was first discovered some 40 years ago [3], and this material is among the most studied cathodic EC materials (darkening under insertion and bleaching under extraction of ions and electrons) [1]. Another, even more widely studied, cathodic EC material is tungsten oxide (WO_3) [1]. Durability under electrochemical cycling is an issue for EC oxides, at least when the goal is to develop commercial products with long lifetime. Addition of titanium has proven to be useful to stabilize the EC properties both for MoO_3 [4] and WO_3 [5].

Research in the field of EC materials recently received new impetus when it was shown that degraded WO_3 thin films can recover their original EC properties by means of driving a constant current, on the order of 10^{-5} A/cm^2, through the films for several hours; this procedure is called *galvanostatic* "rejuvenation" [6]. The surmised reason for the degradation of the films—*viz*., trapped lithium ions—as well as the removal of these ions with galvanostatic treatment were experimentally confirmed by use of Time-of-Flight Elastic Recoil Detection Analysis (ToF-ERDA) applied to as-deposited, degraded and rejuvenated samples [7]. Very recently, *potentiostatic* treatments were performed on WO_3 films, and the possible types and nature of ion traps were elucidated [8]. Furthermore, rejuvenation is feasible via galvanostatic and potentiostatic treatments of titanium oxide (TiO_2), which is another important cathodic EC material [9].

In this present work we prepared MoO_3 thin films by reactive DC sputtering and studied their EC performance using voltammetric cycling in an extended voltage range to ensure degradation within 20 insertion/extraction cycles for ions and electrons. The films were then subjected to potentiostatic treatments with different voltages in order to rejuvenate them, and finally the films were cycled again in a more restricted "normal" voltage range to check their EC performance. The results were compared with data for films cycled using a "normal" voltage range already from the beginning.

2. Thin film deposition

Prior to thin film deposition, the sputter system was evacuated to 2×10^{-4} mTorr. Argon and oxygen, both of 99.997% purity, were then introduced through mass-flow controlled gas inlets. The O_2/Ar gas flow ratio was set to 0.14 and the working pressure was ~30 mTorr. Substrates for optical and electrochemical measurements were unheated 5×5 cm^2 glass plates pre-coated with transparent and electrically conducting layers of In_2O_3:Sn (ITO) having a resistance of 40 Ω/square. The target–substrate separation was 13 cm. Pure MoO_3 films were obtained using a 5-cm-diameter Mo target with 99.95% purity. The discharge power was set to 300 W. Film thicknesses were ~300 nm as determined by surface profilometry using a Bruker DektakXT instrument.

The deposited films displayed X-ray diffraction patterns characteristic of amorphous materials. The films were stored in a dark and dry place for five months before their EC properties were studied.

3. Results and discussion

3.1. Electrochromic performance under a "normal" voltage range

Cyclic voltammetry (CV) measurements were performed using a Solartron 1286 Electrochemical Interface and a three-electrode electrochemical cell. The electrolyte was 1 M $LiClO_4$ in propylene carbonate, and lithium foils served as reference and counter electrodes. All measurements were performed inside a glove box with inert Ar atmosphere. The voltage sweep rate was 10 mV/s.

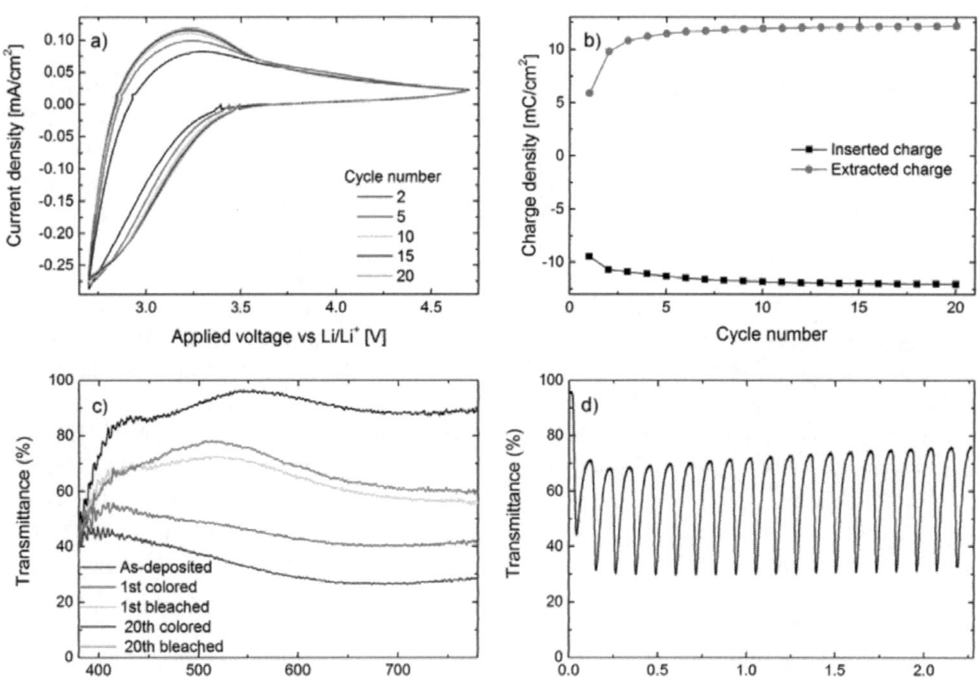

Figure 1. Electrochromic performance of as-deposited MoO_3 films in the voltage range 2.7–4.7 V *vs*. Li/Li^+. (a) Selected cyclic voltammograms. (b) Evolution of the inserted and extracted charge; symbols denoting data were joined by straight lines. (c) Spectral transmittance for the indicated coloured and bleached states. (d) Evolution of the transmittance at the wavelength 550 nm.

Figure 1 shows data for a MoO_3 film cycled in the voltage range 2.7–4.7 V *vs*. Li/Li^+. The first cycles exhibit anomalous data, which is a common feature for many EC oxides, but the following CV cycles tended to remain stable, as shown in panel (a). Figure 1(b) indicates that also the calculated inserted and extracted charge densities were different in absolute value for the first cycles, which indicates that some charge is trapped, but then the charge densities stabilized at similar absolute values. Figure 1(c) reports *in situ* optical transmission, measured with a fiber-optics-based Ocean Optics instrument using an uncoated substrate as a reference, and shows high transparency of the as-deposited films as well as evidence of optical interference. The first bleached state does not reach the initial transparency of the film, which looks slightly bluish in this condition. Nonetheless, the bleached state improves somewhat towards the end of the 20 CV cycles and also the colored state becomes more pronounced at the end of the cycling than for the first cycle. These results are verified by examining the evolution of the monochromatic mid-luminous transmittance at the wavelength of 550 nm, as displayed in figure 1(d). The latter figure also confirms the relative optical stability during the cycling.

3.2. Electrochromic performance under an extended voltage range

Drastic changes appeared when the MoO_3 thin films were cycled within the wider voltage range of 2.2–4.7 V *vs*. Li/Li^+, with all of the other parameters kept the same, as can be seen in figure 2(a). The voltammograms shrank monotonically during the CV cycling, thus indicating that less and less charge was inserted and extracted, as quantified in figure 2(b).

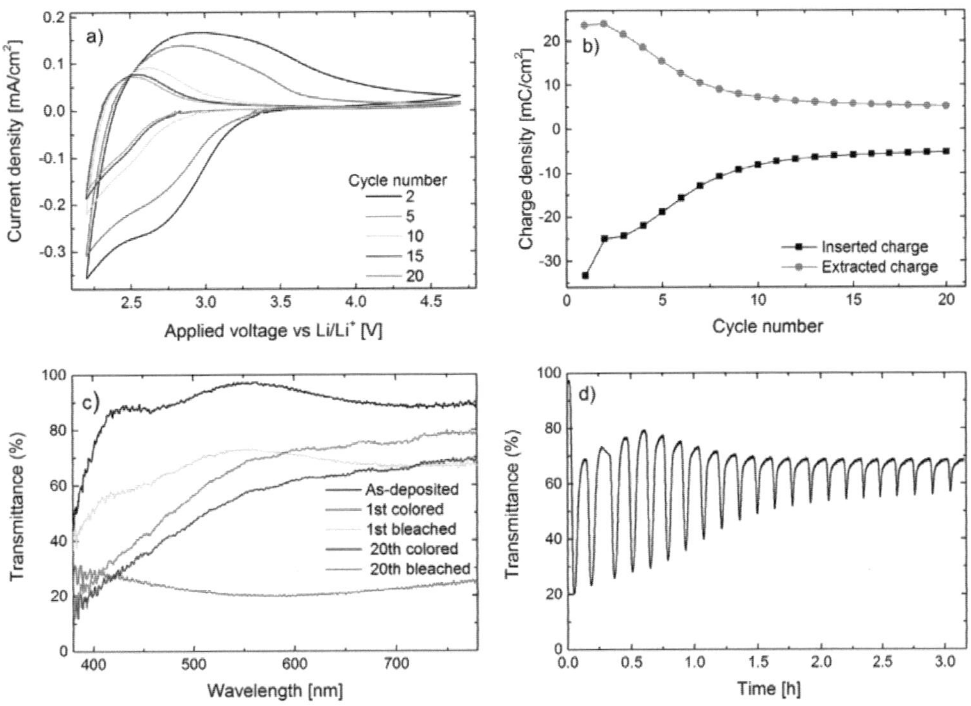

Figure 2. Electrochromic performance of as-deposited MoO_3 films in the voltage range 2.2–4.7 V *vs*. Li/Li^+. (a) Selected cyclic voltammograms. (b) Evolution of the inserted and extracted charge; symbols denoting data were joined by straight lines. (c) Spectral transmittance for the indicated coloured and bleached states. (d) Evolution of the transmittance at the wavelength 550 nm.

As expected, the film was darker in the initial colored state than when the film was cycled at 2.7–4.7 V *vs.* Li/Li^+, as evident from a comparison of data in figures 1(c) and 2(c). Furthermore, the first bleached state was less transparent than the as-deposited film. At the end of the 20 CV cycles, the film acquired a yellowish tint and the EC performance was poor; the transmittance of the bleached and colored states was close. The transmittance modulation at 550 nm was also greatly diminished as the number of CV cycles increased, as apparent in figure 2(d). All of these results are very similar to those observed for EC WO_3 films when cycled in extended voltage ranges [5–7].

3.3. Rejuvenation by potentiostatic treatment

The degraded MoO_3 films were subjected to potentiostatic treatment at a constant voltage for 20 h. The selection of voltages—specifically 5.25, 5.50 and 5.75 V *vs.* Li/Li^+—were guided by recent work on EC WO_3 films [8] and regarding that the "normal" voltage range is 2.0–4.0 V *vs.* Li/Li^+ for the latter films whereas a range of 2.7–4.7 V *vs.* Li/Li^+ is more adequate for MoO_3 films. We also have to consider that higher voltages, exceeding 6.2 V *vs.* Li/Li^+, lead to the risk of altering the electrolyte; in fact, the electrolyte turned slightly yellowish at too large voltages.

Figure 3(a) shows the evolution of the current density during potentiostatic treatments. Despite some noise from the experimental set-up, we could ascertain that a clear difference in the data took place between 5.75 V *vs.* Li/Li^+ and the two lower voltages. Unlike data for WO_3 films, strong maxima were not displayed but only some weak peak-like features after ~11 h for the lower voltages and after ~7 h for 5.75 V *vs.* Li/Li^+. It is worth noting that the feature appeared earlier for the higher voltage, which concurs with results for WO_3 [8].

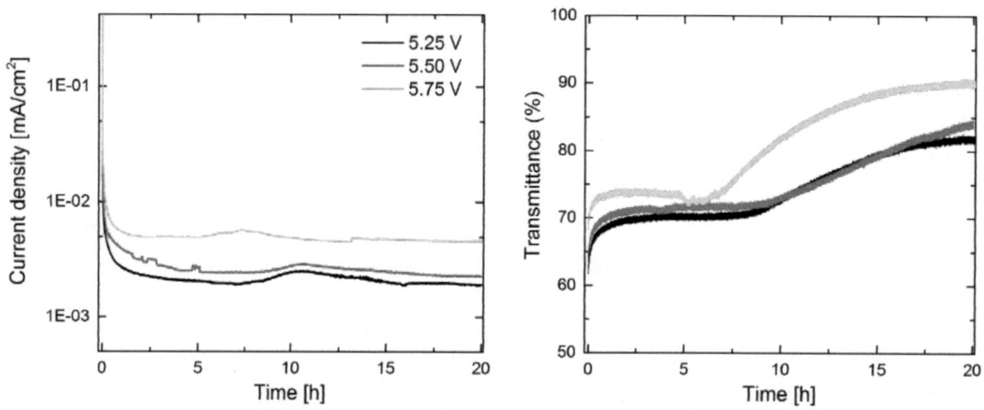

Figure 3. Rejuvenation of degraded MoO_3 films under potentiostatic treatments. (a) Current density and (b) optical transmittance at a wavelength of 550 nm as a function of time at the indicated voltage *vs.* Li/Li^+.

Figure 3(b) displays how the transmittance at 550 nm changes upon various applied voltages. As for the current density, there was little difference when the potentiostatic treatments were performed with 5.25 and 5.50 V *vs.* Li/Li^+: after a first abrupt increase of the transmittance, both curves remained at a constant value for ~8 h and then increased steadily. Intriguingly, the main peak-like features in the current density also took place after ~8 h. For 5.25 V *vs.* Li/Li^+, the transmittance seemed to stabilize at the end of the curve, while for 5.50 V *vs.* Li/Li^+ it appeared that the transmittance could increase even more if the treatment were continued. For 5.75 V *vs.* Li/Li^+, the steady increase of the transmittance occurred somewhat earlier, specifically after ~7 h. The transmittance started to drop somewhat about two hours prior to its main rise which is in principle agreement with observations for WO_3 [8]; the reason for this phenomenon is unknown. The transmittance value in this case seemed to stabilize after 18 h.

Journal of Physics: Conference Series **764** (2016) 012009 doi:10.1088/1742-6596/764/1/012009

3.4. Electrochromic performance after potentiostatic rejuvenation

Once the rejuvenation process was finished, we waited for two hours to let the film relax and attain an open circuit potential similar to that for the as-deposited samples [7]. The films were then cycled again in the range 2.7–4.7 V *vs.* Li/Li$^+$. All films behaved similarly, and we only display data for the sample rejuvenated after potentiostatic treatment at 5.75 V *vs.* Li/Li$^+$. Figure 4(a) shows that the shape and stability within 20 CV cycles resemble what we obtained for the as-deposited samples. Furthermore, the evolution of the inserted and extracted charge density, shown in figure 4(b), was similar to the data in the corresponding figure 1(b); even the quantitative charge densities tended to stabilize at similar values in the two cases. The transmittance spectrum for the rejuvenated film, shown in figure 4(c), was not entirely the same as for the as-deposited sample, which may be due to the fact that some minor amount of lithium remains in low-energy traps in the MoO_3 structure, as previously discussed for WO_3 [6–8], and moreover one cannot discard the possibility that the insertion of lithium modifies the structure of the oxide film to some extent. As noticed before [8], the transmittance of the first bleached state was not as high for the rejuvenated film. Figure 4(d) shows that the improvement in both the last bleached and coloured states, as compared to results for the first state, is even more marked than in the case of the as-deposited sample. Furthermore, figure 4(d) confirms the recovery of the EC properties of the rejuvenated MoO_3 film, and the evolution of the transmittance at 550 nm looks rather stable during the CV cycles. Nonetheless, comparing figures 1(d) and 4(d) reveals that the transmittance modulation is slightly diminished for the rejuvenated film at the end of the 20 CV cycles.

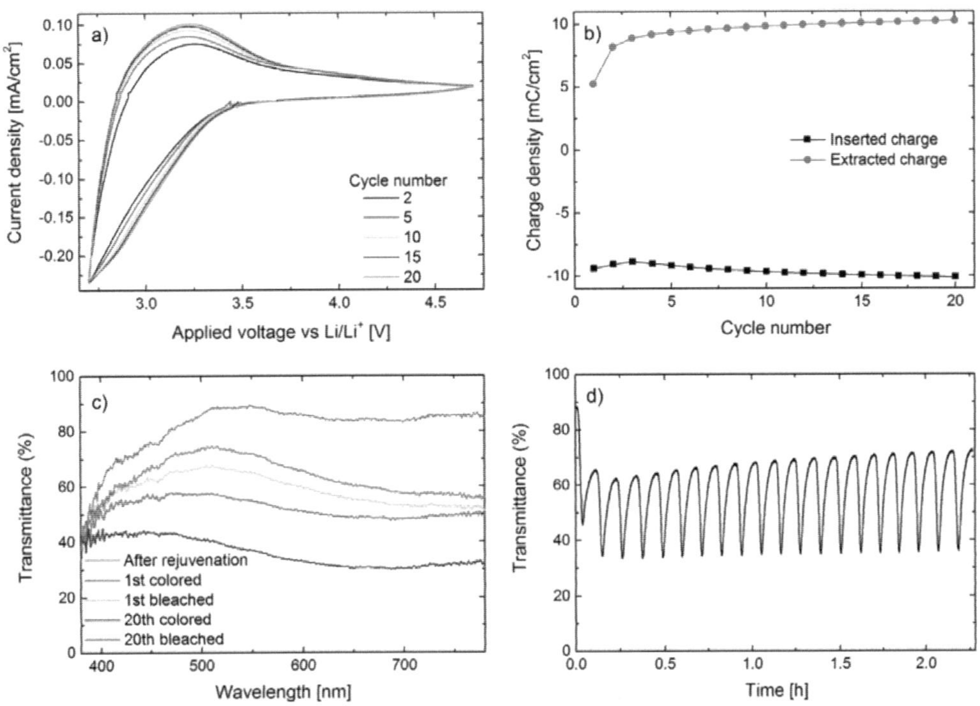

Figure 4. Electrochromic performance of a MoO_3 film after potentiostatic rejuvenation at 5.75 V *vs.* Li/Li$^+$ and in the voltage range 2.7–4.7 V *vs.* Li/Li$^+$. (a) Selected cyclic voltammograms. (b) Evolution of the inserted and extracted charge; symbols denoting data were joined by straight lines. (c) Spectral transmittance for the indicated coloured and bleached states. (d) Evolution of the transmittance at the wavelength 550 nm.

4. Conclusions

We demonstrated that electrochromic MoO_3 films can be potentiostatically rejuvenated after severe degradation upon voltammetric cycling in an extended voltage range. Moreover we were able to establish similarities, as well as some differences, with previous work wherein electrochromic WO_3 films were rejuvenated via similar treatments. We emphasize that the present work is preliminary, and more experiments and analysis are needed to fully elucidate the nature of the rejuvenation of the electrochromic performance.

Acknowledgements

This paper was presented at the INERA Conference on "Vapor Phase Technologies for Metal Oxide and Carbon Nanostructures", July 5–9, 2016, in Velingrad, Bulgaria. The conference is part of the INERA REPGOT project of the Institute of Solid State Physics, Bulgarian Academy of Sciences. MAA thanks the Mexican Council for Science and Technology (CONACyT) for financial assistance to carry out postdoctoral research at Uppsala University. Our work was supported by the European Research Council under the European Community's Seventh Framework Program (FP7/2007–2013)/ERC Grant Agreement No. 267234 ("GRINDOOR").

References

[1] Granqvist C G 1995 *Handbook of Inorganic Electrochromic Materials* (Amsterdam: Elsevier)
[2] Granqvist C G 2014 *Thin Solid Films* **564** 1–38
[3] Deb S K 1975 *Proc. 4th Int. Congr. Reprography and Information* (Hannover, Germany) 115–19
[4] Kullman L, Azens A, Granqvist C G 2000 *Sol. Energy Mater. Sol. Cells* **61** 189–96
[5] Arvizu M A, Triana C A, Stefanov B I, Granqvist C G and Niklasson G A 2014 *Sol. Energy Mater. Sol. Cells* **125** 184–89
[6] Wen R-T, Granqvist C G, Niklasson G A 2015 *Nat. Mater.* **14** 996–1001
[7] Arvizu M A, Wen R-T, Primetzhofer D, Klemberg-Sapieha J E, Martinu L, Niklasson G A, Granqvist C G 2015 *ACS Appl. Mater. Interfaces* **7** 26387–90
[8] Wen R-T, Arvizu M A, Morales-Luna M, Granqvist C G, Niklasson G A 2016 *Chem. Mater.* DOI: 10.1021/acs.chemmater.6b01503
[9] Wen R-T, Niklasson G A, Granqvist C G 2016 *ACS Appl. Mater. Interfaces* **8** 5777–82

Optical, structural and electrochromic properties of sputter-deposited W-Mo oxide thin films

K Gesheva[1,*], M A Arvizu[2], G Bodurov[1], T Ivanova[1], G A Niklasson[2], M Iliev[3], T Vlakhov[3], P Terzijska[3], G Popkirov[1], M Abrashev[4], S Boyadjiev[3,5], G Jágerszki[5], I M Szilágyi[5,6] and Y Marinov[3]
[1]Central Laboratory of Solar Energy and New Energy Sources, Bulgarian Academy of Sciences, 72 Tzarigradsko chaussee, 1784 Sofia, Bulgaria
[2]Department of Engineering Sciences, The Ångström Laboratory, Uppsala University, P.O. Box 534, SE-751 21 Uppsala, Sweden
[3]Institute of Solid State Physics, Bulgarian Academy of Sciences, 72 Tzarigradsko Chaussee blvd., 1784 Sofia, Bulgaria
[4]Faculty of Physics, Blvd. "J. Bauchier" 5, Sofia University, Bulgaria
[5]MTA-BME Technical Analytical Chemistry Research Group, Szent Gellért tér 4, Budapest, H-1111, Hungary
[6]Department of Inorganic and Analytical Chemistry at Budapest University of Technology and Economic, Szent Gellért tér 4, Budapest, H-1111, Hungary

E-mail: kagesh@phys.bas.bg

Abstract. Thin metal oxide films were investigated by a series of characterization techniques including impedance spectroscopy, spectroscopic ellipsometry, Raman spectroscopy, and Atomic Force Microscopy. Thin film deposition by reactive DC magnetron sputtering was performed at the Ångström Laboratory. W and Mo targets (5 cm diameter) and various oxygen gas flows were employed to prepare samples with different properties, whereas the gas pressure was kept constant at about 30 mTorr. The substrates were 5×5 cm^2 plates of unheated glass pre-coated with ITO having a resistance of 40 ohm/sq. Film thicknesses were around 300 nm as determined by surface profilometry. Newly acquired equipment was used to study optical spectra, optoelectronic properties, and film structure. Films of WO_3 and of mixed W–Mo oxide with three compositions showed coloring and bleaching under the application of a small voltage. Cyclic voltammograms were recorded with a scan rate of 5 mV s^{-1}. Ellipsometric data for the optical constants show dependence on the amount of MoOx in the chemical composition. Single MoOx film, and the mixed one with only 8% MoOx have the highest value of refractive index, and similar dispersion in the visible spectral range. Raman spectra displayed strong lines at wavenumbers between 780 cm^{-1} and 950 cm^{-1} related to stretching vibrations of WO_3, and MoO_3. AFM gave evidence for domains of different composition in mixed W-Mo oxide films.

1. Introduction

Smart windows allow control of the solar flux entering buildings. The functional layer, a transition metal oxide is, because of its specific electronic structure capable to change the transmittance if an electrical field is applied across it. It switches between transparent and coloured state, thus controlling the visible and near-infrared solar radiation entering the buildings or vehicles. In order to show up its ability to change the optical transmittance in practical applications, the layer should be part of a multilayered system. Some transition metal oxides exhibit electrochromic properties, when employed as part of an electrochromic (EC) multilayer system – most common is the so-called smart window. Such a system (figure 1) comprises several layers in a multilayer stack – float glass; transparent conducting oxide (TCO - mostly SnO_2, tin doped indium oxide (ITO), fluorine doped tin oxide (FTO), Al-doped zinc oxide (AZO) or a polymer conductor); functional EC electrode (mostly tungsten trioxide WO_3); ion conducting material, with absent or negligibly low electron conductivity; ion storage electrode (another EC material); and second transparent conductor [1].

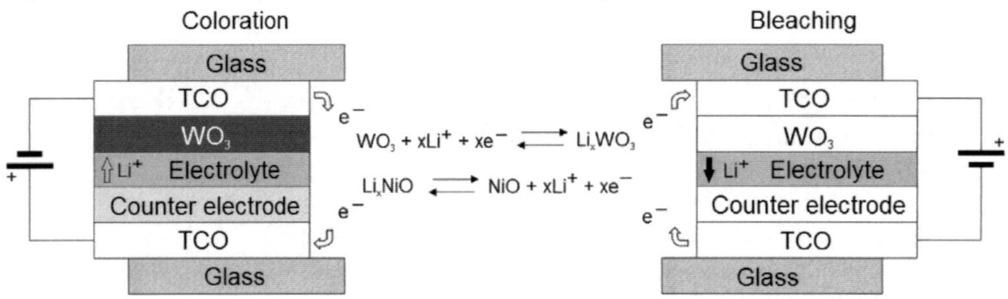

Figure 1. Schematic representation of an EC multilayer stack.

WO_3 thin films are the most widely studied EC oxides, which colour by reduction (cathodic coloration), and have been fabricated by a variety of coating methodologies. For achieving flow-through processes in the manufacturing phase, and for economic reasons, low-cost and large-scale methods are preferable. Tungsten trioxide films have been prepared by different deposition techniques [1-4], including vacuum evaporation, electrochemical deposition, chemical vapour deposition and sputtering. Non-vacuum approaches such as spray pyrolysis (SP) and sol-gel deposition are considered as low cost alternatives [5-11]. Molybdenum oxide (MoO_3) is another cathodic EC material, attractive because of the position of its optical absorption peak, which is near the human eye sensitivity peak. This work presents preliminary results on the characterization of sputtered as-deposited MoO_3, WO_3 and mixed Mo-W oxide films. Newly acquired equipment was used to study optical spectra, optoelectronic properties and film structure. Films of WO_3 and of mixed W–Mo based oxides with three compositions showed colouring and bleaching under application of a small voltage, but in this initial report we mainly present results pertaining to the bleached state.

2. Experimental

Thin film deposition by reactive DC magnetron sputtering utilized W and Mo targets with 5 cm diameter and various oxygen gas flows were employed to prepare samples with different properties, whereas the gas pressure was kept constant at about 30 mTorr. The substrates were 5×5 cm^2 plates of unheated glass pre-coated with ITO having a resistance of 40 ohm/sq. The detailed deposition conditions were the same as in a previous study of W-, Mo- and mixed W-Mo oxide films [12]. Film thicknesses for all the samples were ~300 nm as determined by surface profilometry using a Bruker DektakXT instrument. X-ray diffraction measurements have been reported before [12] and show that the samples were X-ray amorphous.

The Raman spectra were obtained using LabRAM HR Visible micro-Raman spectrometer. The excitation light was the 633 nm line of a He-Ne laser. An X100 objective focused the incident beam to a spot with a diameter of about 1-2 μm and collected the scattered light in the backscattering configuration. No overheating effects were observed at the laser power used (5.7 mW on the laser spot). AFM observations were performed with a Nanosurf FlexAFM in the Department of Inorganic and Analytical Chemistry at Budapest University of Technology and Economics at the following conditions: tapping mode; Tap300GD-G cantilever; vibration amplitude = 100mV.We note that all the measurements were performed for the films in normal, uncoloured state.

Ellipsometry measurements were performed using a J.A. Woollam Co., Inc. M2000D rotating compensator spectroscopic ellipsometer with a CCD spectrometer with wavelength range from 193 to 1000 nm. Experimental data for the ellipsometric angles Ψ and Δ were acquired at angles of incidence of 65, 70 and 75 degrees. The data were modelled using the CompleteEASE Woollam Co., Inc. software, in order to obtain the refractive index and extinction coefficient of the samples. Cyclic voltammetry (CV) and impedance spectroscopy was performed by a Bio-logic SP-200 potentiostat using a standard three-electrode configuration. The electrochromic film was set as working electrode, a 1 cm^2 platinum plate as counter electrode, and a Standard Calomel Electrode (SCE) was used as reference electrode. The electrodes were immersed in 1M electrolyte of lithium perchlorate in propylene carbonate. All the measurements were carried out at room temperature. Cyclic voltammetry was performed at voltage sweep rate of 5 mV/s. Impedance spectroscopy was measured in the frequency range from 50 mHz to 1 MHz and using voltages of 10 mV vs. SCE.

3. Results and discussions

Comparison of Raman spectra of the sputtered metal oxide films are presented in figure 2 and Raman bands and their assighments are given in table 1. Raman spectra of the three mixed oxide samples $W_{0.92}Mo_{0.08}O_3$, $W_{0.86}Mo_{0.14}O_3$ and $W_{0.7}Mo_{0.3}O_3$ show Raman peaks that coincide, figure 2 (right side).

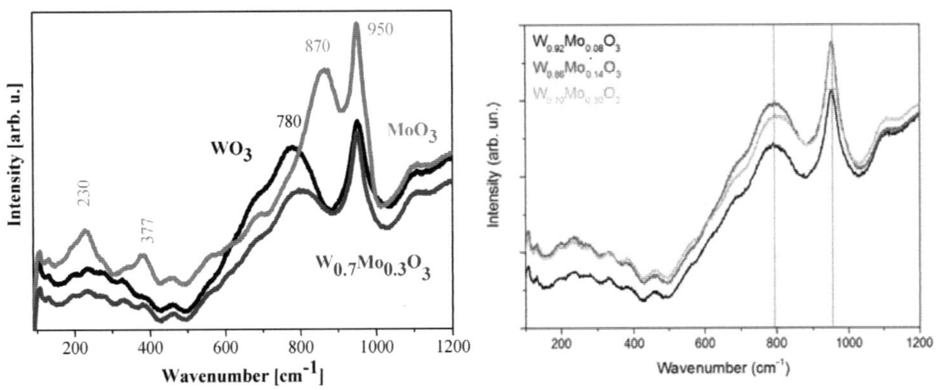

Figure 2. Raman spectra of magnetron sputtered metal oxide films, single W and Mo oxides and a mixed oxide (left) and comparison of mixed W-Mo oxide magnetron sputtered films (right).

The spectrum of MoO_3 differs from the other spectra of WO_3 and mixed metal oxide films. The Raman spectra of WO_3 and the mixed films show similar behaviour. Their Raman spectrum displays strong lines at 780 cm^{-1} and 950 cm^{-1}, attributed to the stretching vibrations of WO_3. The band at 950 cm^{-1} for mixed oxides has a possible contribution from W=O and Mo=O double bonds. The broadness of the bands is a sign that the films are very disordered. Previous studies of WO_3, MoO_3 and mixed oxide films, obtained by APCVD technology, employing carbonyl process [13] have revealed similar features, namely a similarity of the Raman spectra of single WO_3 film with the W-Mo-O mixed films.

The AFM images show that the film surfaces are characterized by irregularly distributed domed crystallites.The images of 2D and 3D topography are presented in the figure 3 below. In figure 3c, where the 3D topography of the mixed $W_{0.92}$ $Mo_{0.08}$ O_3 film is presented, the phase differences are marked in colors. Phase contrast imaging allows us to detect regions with different compositions on the film surface. The larger region (dark/blue) we suggest to relate to WO_3, and the bright/yellow-greenish colors to MoO_x. The small amount, only 8 % of Mo in the mixed film supports that conclusion. This technique thus shows that the mixed W-Mo oxide films exhibits regions of different composition, which significantly adds to our knowledge of the structure of those films.

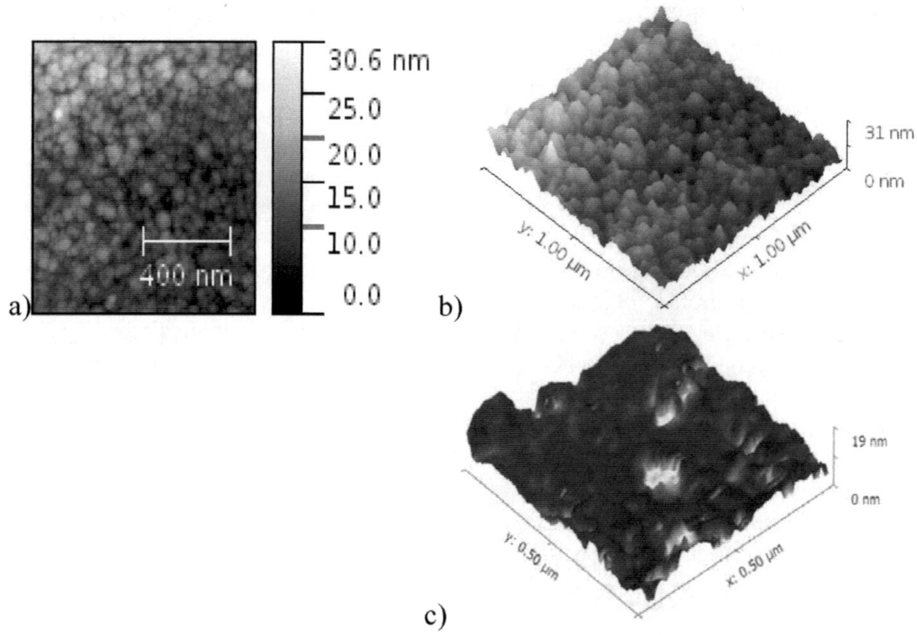

Figure 3. AFM 2D (a), and 3D (b) topography of sputtered WO_3 thin film, and AFM 3D topography together with phase difference (in colour) of magnetron sputtered mixed $W_{0.92}Mo_{0.08}O_3$ thin film (c).

Impedance spectroscopy results are presented in the Nyquist complex impedance plot, figure 4. It shows the imaginary versus the real part of electrical impedance for a tungsten oxide and a mixed W-Mo oxide film at a potential of -0.4V vs. SCE.

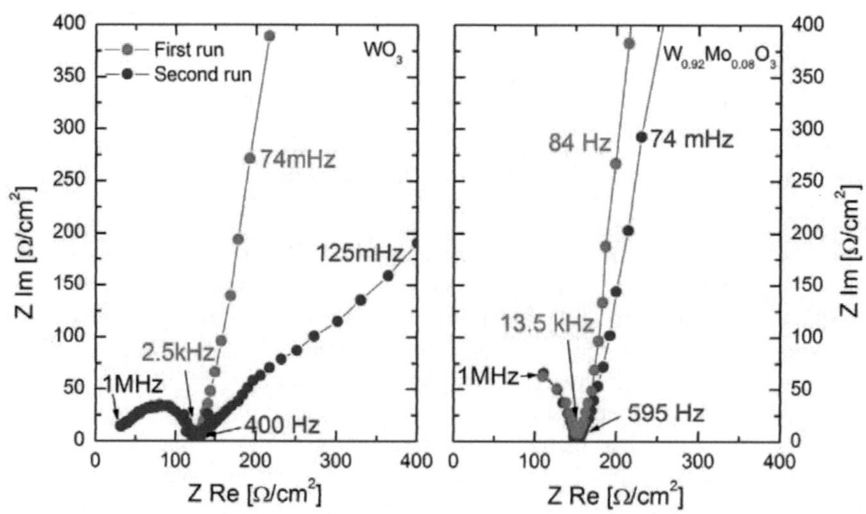

Figure 4. Nyquist complex impedance plot showing the imaginary part of electrical impedance (ZIm) as a function of the real part (ZRe), for a tungsten oxide and a mixed film at -0.4V dc potential vs. SCE. The frequencies of measured points that outline the high and low frequency regions are indicated.

Journal of Physics: Conference Series **764** (2016) 012010 doi:10.1088/1742-6596/764/1/012010

It is seen that the response in the high frequency region is represented by a depressed semicircle while in low frequency region it approximates a straight line. The 'semicircle' is related to the electric double layer formed at working electrode and the line corresponds to the diffusion process of ions into the tungsten layer. The obtained impedance spectra are typical for tungsten oxide films [14] and can probably be modelled by a Randles circuit consisting of a charge transfer resistance, a double layer constant phase element and anomalous diffusion elements [15]. In comparison to WO_3, the mixed W/Mo thin film shows higher double layer relaxation frequency (above 1 MHz) and higher starting frequency of the diffusion region. The mixed films demonstrate higher value of the charge transfer resistance, 152 vs. 127 ohms/cm^2 for WO_3.

Results shown on figure 4 also show the result of a repeated measurement and it is seen that the progress of ion insertion into the film for both cases leads to a decrease of the diffusion line slope. The mixed film appears to have a better stability at this potential than the WO_3 film.

Voltammograms for the magnetron sputtered films were measured with a scan-rate of 5mV/cm^2. As seen from figure 5, voltammogram cycles of WO_3 thin films have shown a good repeatability, observed in the selected interval of voltages. For the mixed W/Mo based magnetron sputtered film a slight difference is seen between the voltammograms.

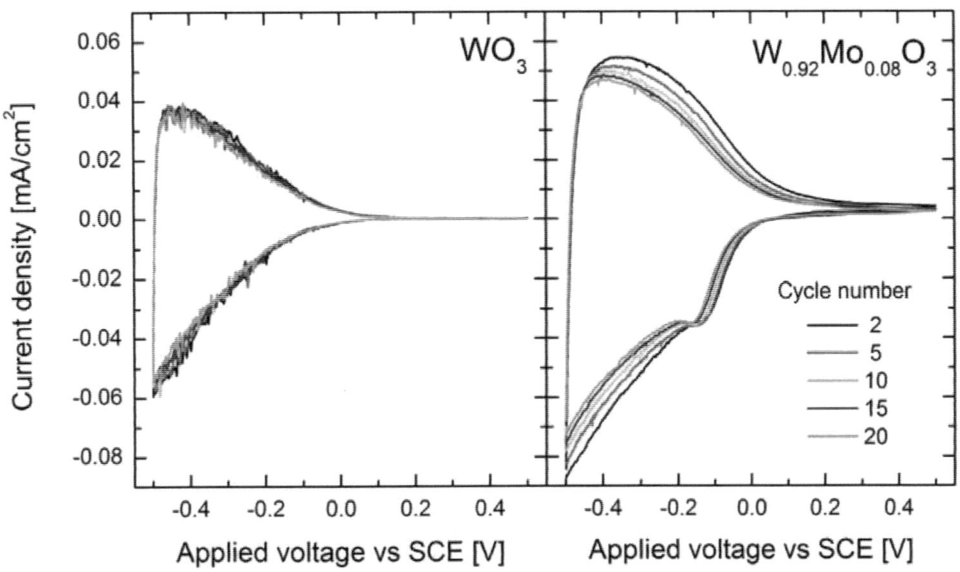

Figure 5. Cyclic voltammograms for WO_3 (left) and mixed Mo-W-O film (right) immersed in 1 M $LiClO_4$–PC electrolyte.

In figure 6 the calculated charge density is shown for pure tungsten oxide and mixed W-Mo oxide corresponding to the voltammograms in figure 5. As in the previous cited work [12], the charge density for the mixed oxide is larger than for pure tungsten oxide. Also the presence of Mo seems to have a detrimental effect in the durability of the film since the charge density is more stable for pure tungsten oxide.

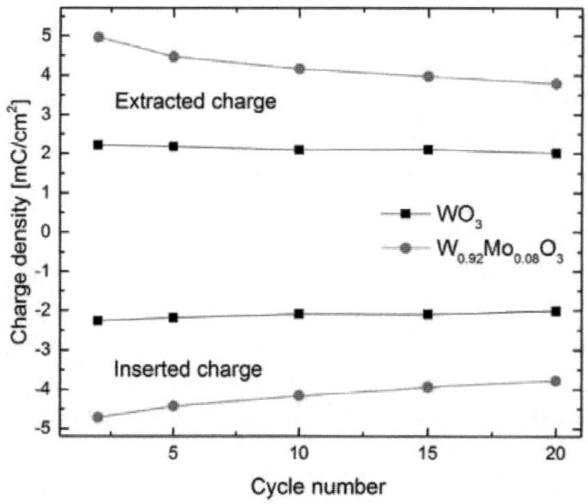

Figure 6. Inserted and extracted charge for the indicated type of films as a function of cycle number. Symbols denoting data are joined by straight lines.

Ellipsometric measurements of all the metal oxide films obtained by magnetron sputtering were performed. Film thickness values, determined by profilometry have been estimated to be around 300 nm, and by ellipsometry measurements the film thicknesses were further precisely determined. The film thickness of MoO_3 and WO_3 films are 305 nm and 310.8 nm, respectively. The $W_{0.92}Mo_{0.08}$ O_3 film is 287.5 nm thick. With increasing of Mo content, the film thickness increases to 306 nm for the $W_{0.7}$ $Mo_{0.3}$ O_3, and to 309.8 nm for $W_{0.86}Mo_{0.14}O_3$. The values of films thickness differ slightly, the difference does not exceed 22 nm.

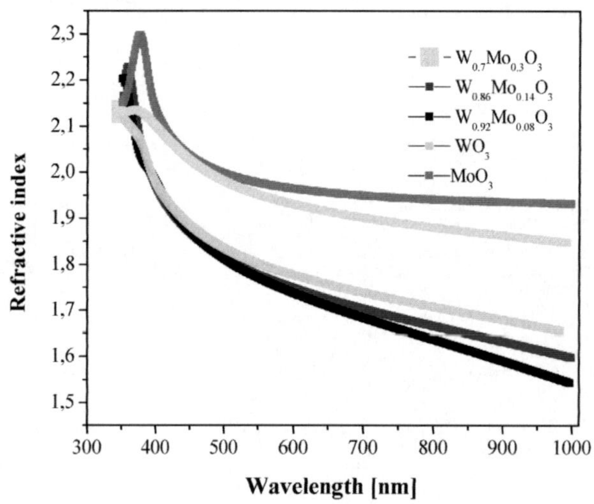

Figure 7. Refractive index as a function of wavelength for the magnetron sputtered single WO_3, MoO_3 and the mixed oxide films.

The results derived from the ellipsometric mesurements for the optical constants, refractive index and extinction coefficient, of the single metal oxide films, and the three mixed films are presented in figures 7 and 8.

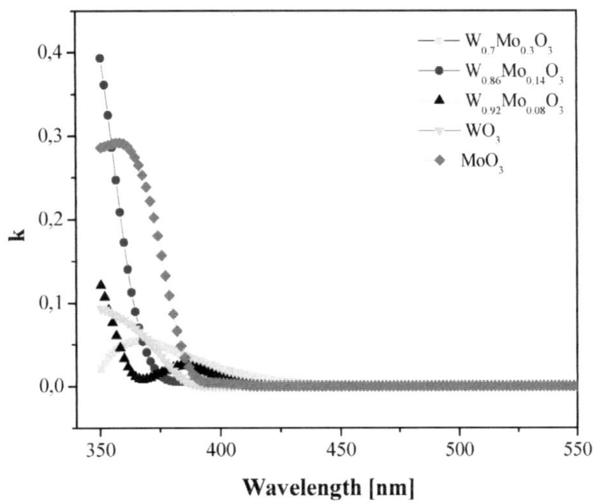

Figure 8. Extinction coefficient as a function of wavelength for the magnetron sputtered WO_3, MoO_3, and for different compositions of mixed oxide films.

The refractive index for MoO_x films has the highest values, and the mixed oxide film $W_{0.7}Mo_{0.3}O_3$ has refractive index values close to the ones of MoO_x. The other two mixed compositions where the amount of MoO_x is much smaller, have refractive index dispersion similar to the one of WO_3 in the spectral range 300 - 600 nm and a minor difference is observed for wavelengths above 600 nm.

The extinction coefficients for all samples are very low, the values are approaching zero at wavelengths longer than 400 nm. The range of wavelengths where a sharp change of the coefficient of extinction appears, see figure 7, is 340 - 390 nm. This means that band gap energies are above 3 eV. The transition metal oxides are wide band-gap semiconductors. It is known that the optical band gap of amorphous WO_3 is in the range 3.1 to 3.2 eV [16]. The optical band gap of sputtered MoO_3 is reported by other authors [17] to be above 3.1 eV depending on the oxygen partial pressure during film deposition. Data for band gap values of APCVD WO_3, MoO_3 and MoO_3/WO_3 are in the same range [13]. The $W_{0.92}Mo_{0.08}O_3$ and $W_{0.7}Mo_{0.3}O_3$ films show significant absorption extending also above 400 nm. This may be due to a lower band gap for these films or alternatively they exhibit an absorption tail due to disorder extending towards the visible range.

4. Conclusions

Structural, optical and electrochemical properties were studied for magnetron sputtered transition metal oxide thin films, namely MoO_3 and WO_3 films, and three selected chemical compositions of mixed films. The properties for the mixed films are influenced by the chemical composition, more exactly by the amount of MoOx component. Raman spectroscopy and AFM were used to characterize the films. AFM gave evidence for domains of different compositions in the mixed films. The refractive index and extinction coefficient, show strong disperson in the studied wavelength range. The value of the refractive index of MoOx is higher than the value for WO_3, while the refractive index of $W_{0.7}Mo_{0.3}O_3$ film approaches that of MoO_3. The band gap is above 3 eV, except for some of the mixed films. Impedance spectra of the mixed W/Mo thin film show higher relaxation frequency (above 1 MHz) and higher starting frequency of the diffusion region. In addition, the mixed film demonstrates higher value of the charge transfer resistance. The progress of ion insertion into the film for both cases leads to differences in the diffusion process.

Acknowledgements
The authors acknowledge INERA "Research and Innovation Capacity Strengthening of ISSP BAS in Multifunctional Nanostructures" Project support for providing new characterization equipment as well as for facilitating mobility exchange visits between the Ångström Laboratory in Sweden and the ISSP-BAS and its partner institution, the CL SENES, in Bulgaria. Additional support for work at the Ångström Laboratory was supplied by the European Research Council under the European Community's FP7 Grant Agreement No. 267234 ("GRINDOOR")

References

[1] Rao M C and Hussain O M 2011 "Optical properties of vacuum evaporated WO_3 thin films" *Res. J. Chem. Sci.* **2231** 606X

[2] More A J, Patil R S, Dalavi D S, Mali S S, Hong C K, Gang M G, Kim J H and Patil P S 2014. "Electrodeposition of nano-granular tungsten oxide thin films for smart window application. *Mater. Lett.* **134** 98-301

[3] Granqvist C G 2014 "Electrochromics for smart windows: Oxide-based thin films and devices" *Thin Solid Films* **564** 1–38

[4] Wang C K, Lin C K, Wu C L, Brahma S, Wang S C and Huang J L 2013 "Characterization of electrochromic tungsten oxide film from electrochemical anodized RF-sputtered tungsten films" *Ceramics Inter.* **39** 4293-4298

[5] Wei H, Xingru, Y, Qiang W, Shijie W, Yuanbing M, Zhiping L, Haoran C, Luyi S, Suying W and Zhanhu G 2013. "Electrochemical properties and electrochromic behaviors of the sol–gel derived tungsten trioxide thin films." *Energy Envir. Focus* **2** 112-120

[6] Patil C E, Tarwal N L, Jadhav P R, Shinde P S, Deshmukh H P, Karanjkar M M, Moholkar A V, Gang M G, Kim J H and Patil P S 2014 "Electrochromic performance of the mixed V_2O_5-WO_3 thin films synthesized by pulsed spray pyrolysis technique" *Current Appl. Phys.* **14** 389-395

[7] Mukherjee R, Prajapati C S and Sahay P P 2014 "Tin-incorporation induced changes in the microstructural, optical and electrical behavior of tungsten oxide nanocrystalline thin films grown via spray pyrolysis" *J. Thermal Spray Techn.* **23** 1445-1455

[8] Denayer J, Aubry P, Bister G, Spronck G, Colson P, Vertruyen B, Lardot V, Cambier F, Henrist C and Cloots R 2014 "Improved coloration contrast and electrochromic efficiency of tungsten oxide films thanks to a surfactant-assisted ultrasonic spray pyrolysis process" *Solar Energy Mater. Solar Cells* **130** 623-628

[9] Costa C, Pinheiro C, Henriques I and Laia C A 2012 "Inkjet printing of sol–gel synthesized hydrated tungsten oxide nanoparticles for flexible electrochromic devices" *ACS Appl. Mater. Interfaces* **4** 1330-1340

[10] Georg A, Graf W and Wittwer V 2008 "Switchable windows with tungsten oxide." *Vacuum* **82** 730–735

[11] Medina A, Solis J L, Rodriguez J and Estrada W 2003 "Synthesis and characterization of rough electrochromic phosphotungstic acid films obtained by spray-gel process" *Sol. Energy Mater. Sol. Cells* **80** 563–576

[12] Arvizu M A, Granqvist C G, Niklasson G A 2016 "Electrochromism in sputter deposited $W_{1-y}Mo_yO_3$ thin films" *Journal of Physics: Conference Series* **682** 012005

[13] Gesheva K, Szekeres A and Ivanova T 2003 "Optical properties of chemical vapor deposited thin films of molybdenum and tungsten based metal oxides" *Sol. Energy Mater. Sol. Cells* **76** 473–481

[14] Niklasson G A, Malmgren S and Strömme M 2016 "Impedance response of electrochromic materials and devices" in Barsoukov Y. and Macdonald J R (Eds.) *Impedance spectroscopy*, to be published.

[15] Bisquert J and Compte A 2001 "Theory of the electrochemical impedance of anomalous diffusion." *J. Electroanal. Chem.* **499** 112-120

[16] Triana C A, Granqvist C G and Niklasson G A 2015 "Electrochromism and small-polaron hopping in oxygen deficient and lithium intercalated amorphous tungsten oxide films" *J. Appl. Phys.* **118** 024901

[17] Subbarayadu S, Madhavi V and Uthanna S 2013 "Growth of MoO_3 Films by RF Magnetron Sputtering: Studies on the Structural, Optical, and Electrochromic Properties" ISRN Condensed Matter Physics 2013, Article ID 806374

Journal of Physics: Conference Series **764** (2016) 012011 doi:10.1088/1742-6596/764/1/012011

Thermal treatment of solution–processed nano-sized thin films of molybdenum oxide

M Ganchev*, M Sendova-Vassileva, G Popkirov and P Vitanov

Central Laboratory of Solar Energy & New Energy Sources – Bulgarian Academy of Sciences, 72 Tzarigradsko chaussee, 1784 Sofia

*email: mganchev@abv.bg

Abstract. A solution based deposition method to form nano-sized thin films of molybdenum oxide suitable for photovoltaic device applications is presented. The samples were deposited by spin-coating from molybdenum metal organic precursor solution on soda lime glass substrates. The influence of the process parameters such as spinning regime and concentration of the precursor solutions on the thickness and morphology of the films were investigated. The thermal decomposition of the molybdenum precursor and oxide formation were investigated by differential scanning calorimetry and characteristic patterns showed transitions up to 300°C followed by a zone of stability. Optical spectroscopy measurements in the wavelength range from 300 to 1800 nm presented an increase in transparency when temperature of annealing was raised up to 400°C. Raman scattering analysis revealed the presence of mixed molybdenum oxides. Measurements of the electrical conductivity were performed by the 4-point method.

1. Introduction

Transition metal oxides present a large material basis for research and development in a variety of opto-electronic devices. They possess wide possibilities for tuning their electronic properties and an intrinsic chemical stability against oxidation which makes them attractive for applications in organic electronics such as light emitting diodes (OLED), liquid crystal displays (OLCD), thin film transistors (OTFT) and photovoltaics (OPV) [1]. The perovskite solar cells as devices of particular interest noted a remarkable progress and over the last three years have boosted power conversion efficiency from approximately 10% in 2012 [2] to over 21% today [3]. Similar performance brings these devices very close to the efficiencies of leading commercial technologies. Perovskite solar cells are built in two typical configurations: on meso-structured scaffold wide-band gap oxide semiconductors [3, 4] and as planar construction of thin films on a transparent substrate. The planar configuration, developed as adaptation of concept for polymer organic photovoltaics, includes perovskite absorber film sandwiched between hole and electron-transporting layers, processed by simple solution technologies and without high-temperature treatments and annealing [5]. Very often poly (3, 4-ethilene dioxythiophene): poly (styrene sulfonate) – PEDOT:PSS is used as hole-transport layer. Deposited onto the ITO substrate it has favorable influence on the wettability of the coating solution of the halide perovskite. Later it was taken advantage by this low temperature technology to process perovskite solar cells on flexible substrates [6]. However, the PEDOT: PSS contains an active sulphonic group which can lead to corrosion of the ITO substrate or attack and destroy the perovskite, resulting in deteriorated properties of the devices in the long-term [7]. Several alternatives for PEDOT:PSS were attempted, such as NiO_x [8], graphene oxide [7] and PbS [9].

MoO_3 appears to be a prospective hole-transporting material due to its high general stability, proper electronic properties and non-toxic behavior. Solution processed MoO_x was successfully used in efficient organic polymer photovoltaics structures where a remarkable electrical stability was observed [10]. Occasionally, non-ideal current – voltage characteristics are noticed and more detailed information for electro - physical properties of the structure could be accessed by additional measurements in dark and light conditions [11]. Transport properties and recombination phenomena could be evaluated by investigation of transient processes [12]. Comprehensive research in solution processing and thin films MoOx characterization were performed in [13] and transient dynamics in bulk hetero-junction solar cells with the related material were presented in [14]. In this work we have investigated a solution process for deposition of thin films of MoO_x and evaluated optical, structural and morphological properties after annealing at different temperatures in inert atmosphere.

2. Experimental

Thin films MoO_x were deposited by spin coating at 3000 min^{-1} for 30 sec of precursor solutions of molybdenum (VI) oxide bis (2,4 – pentanedionate) - $MoO_2(AcAc)_2$ (mp 184°C) in absolute ethanol on soda lime glass substrates. Subsequently the samples were annealed at different temperatures in a nitrogen glove box. The deposition stock solution was 184 mM (60 g/l) (M1). For deposition of thinner films diluted blends in the ratio 1:2 (92 mM - M2 (30 g/l)) were used.

Mettler DSC 20 standard cell with TC-10A processor, connected to a PC was used for differential scanning calorimetry. Measurements of the optical transmission and reflection spectra were performed in the interval of 300 – 1800 nm using a Shimadzu UV – 3600 spectrophotometer. Raman spectra excited by a HeNe laser were obtained using a Horiba Jobin Yvon LabRam HR 800 spectrometer.

AFM measurements were carried out in tapping mode using a Multimode V unit by Bruker, Santa Barbara, CA. The image processing was performed using SPIP^{tm} 6.1.0 software. The film thickness and spectral distribution of the refractive index and the extinction coefficient were done on an automatic spectroscopic ellipsometer - M2000D, Woollam.Santa Barbara, CA).

3. Results and discussion

3.1. Structural characterization

Differential Scanning Calorimetry analysis (figure 1) was performed in order to gain an insight of the thermal limits and details of the phase transitions during annealing of the thin films deposited from precursor solutions of molybdenyl acetylacetonate $MoO_2(AcAc)_2$ in absolute ethanol. As it is well known, the melting point (accompanied with decomposition) of the compound is 184°C. Curve (1) was obtained for powder of $MoO_2(AcAc)_2$ and presents two endothermic transitions - the first being at 198°C and the second at 247°C. They could be attributed to the decomposition of $MoO_2(AcAc)_2$ and oxide formation, respectively. A subsequent measurement of the same probe (not shown in Fig.1) had no characteristic features, indicating the irreversibility of the reactions. A sample of $MoO_2(AcAc)_2$ which remained overnight under atmosphere of saturated ethanol vapors in nitrogen atmosphere presented two new peaks, at 98°C and 296°C respectively (curve 2) additional to the transitions at 198°C and 248°C for molybdenyl acetylacetonate (cf. curve 1). The first one at 98°C could be assigned to ethanol releasing while the last one at 296°C denotes the point of cleavage of molybdenum - alcohol bonds [15]. Curves 3 and 4 present DSC data of suspensions of $MoO_2(AcAc)_2$ and absolute ethanol in weight ratio 50:50 and 30:70, respectively. The characteristic endothermic transitions at 98°C, 198°C and 247°C are seen again, but at 298°C an exothermic signal appears which should be assigned to the firing of the solvent. Curve (5) obtained for the solution of 184 mM $MoO_2(AcAc)_2$ in absolute ethanol presented the main transition at 98°C whereas the others are negligible probably because of lower concentration. The presented DSC analysis of the active compound and precursor solution points out the characteristic phase transitions and their respective temperature ranges and could help us to determine the desired regime for thermal annealing.

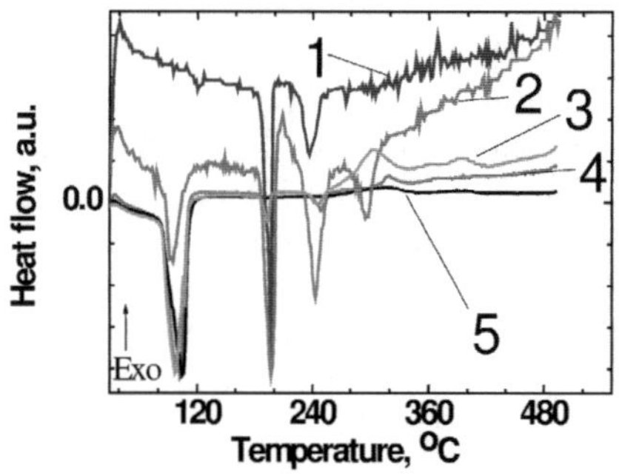

Figure 1. Differential scanning calorimetry: (1) Dry $MoO_2(AcAc)_2$; (2) $MoO_2(AcAc)_2$ under ethanol vapors for 24 hours; (3) 1:1 by weight mix of $MoO_2(AcAc)_2$ and ethanol; (4) 1:3 by weight mix of $MoO_2(AcAc)_2$ and ethanol; (5) Solution of 184 mM $MoO_2(AcAc)_2$ in ethanol.

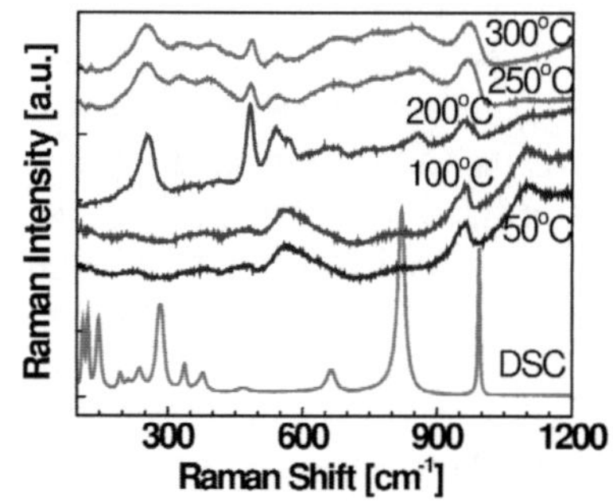

Figure 2. Normalized Raman spectra of MoO_x thin films, deposited with solution M1and annealed in nitrogen at different temperatures shown in the figure as well as of the powder material left after the DSC measurement.

The Raman spectra of the MoO_x thin films, deposited with solution M1 and annealed at different temperatures are shown in figure 2. The spectrum of the powder left after the DSC experiment is shown for comparison. That is a typical spectrum of orthorhombic MoO_3 with clearly visible characteristic bands - 996 cm^{-1} Mo=O stretching, 820 cm^{-1} Mo-O-Mo stretching, 666 cm^{-1} O-Mo stretching, deformation modes below 400 cm^{-1} [16]. The Raman spectra of the films annealed at increasing temperatures exhibit a number of broad bands which could be connected to the evolving disordered molybdenum oxide matrix. At 50ºC and 100ºC only bands in the range 960 – 1100 cm^{-1} and around 560 are observed which could tentatively be assigned to vibrations of Mo=O and Mo-O bonds, respectively, of oxygen atoms bound to a single Mo metal ion [17].

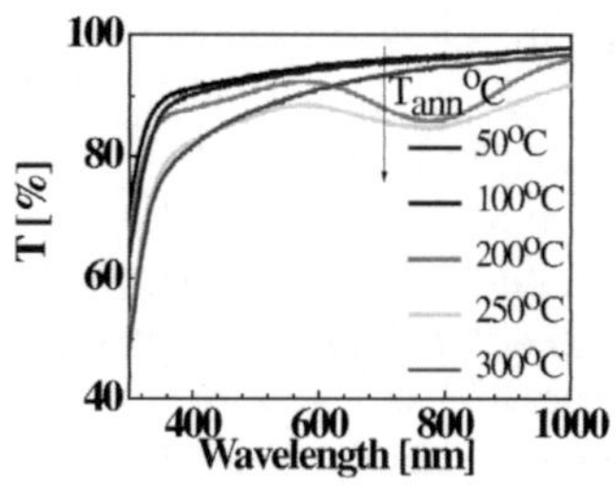

Figure 3. Optical transmission of MoO_x thin films, deposited with solution **M1** and annealed at different temperatures in nitrogen.

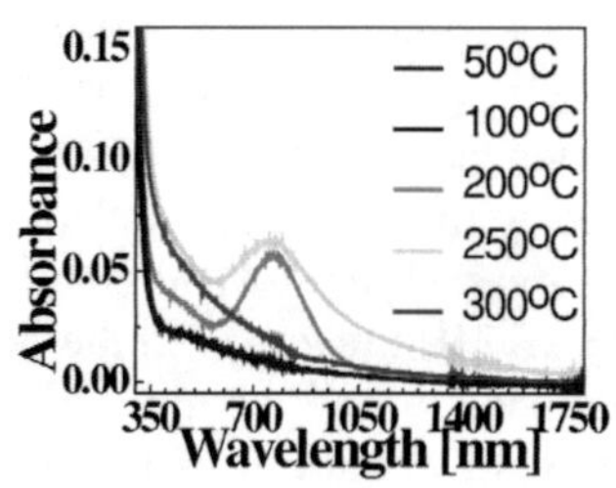

Figure 4. Absorbance of MoO_x thin layers, deposited with solution **M1**and annealed at corresponding temperature in nitrogen.

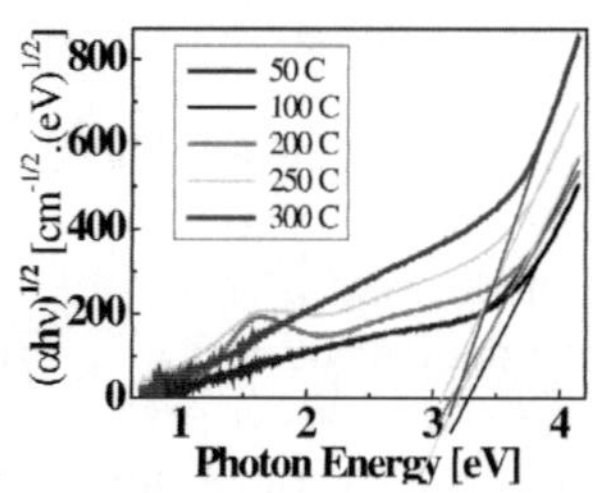

Figure 5. Tauc plot for determination of the band gap of disordered MoO_x thin films.

With increasing temperature bands at around 860 cm^{-1} and one around 255 cm^{-1} appear. The first one could be connected with Mo-O-Mo bonding and the second one to deformational vibrations. The bands between 480 and 580 cm^{-1} acquire more structure at 200ºC. This evolution is evidence of increased structural order and interconnectedness of the MoO_x matrix.

Journal of Physics: Conference Series **764** (2016) 012011 doi:10.1088/1742-6596/764/1/012011

3.2. Optical properties

The deposited thin films demonstrate high transparency (figure 3) which is of great importance for application in solar cells. As shown in the figure the transparency falls a little with the temperature of annealing for the used composition of the precursor solution. After annealing at 200ºC there appears a well formed absorption band at 770 nm (1.6 eV) (cf. figure 4) which is gradually annealed out at higher temperatures. This trend was observed for evaporated layers of MoO_x and the effect is explained with an unstable equilibrium in phase transitions between oxides of Mo^{5+} and Mo^{6+}in this temperature interval[18,19] Figure 5 shows the Tauc plot for assessing the optical band gap of the disordered oxide films. The values obtained are compatible with the literature values for MoO_x thin films.

Figure 6. AFM picture of a MoO_x thin film, deposited with solution M1, annealed at 300°C in nitrogen.

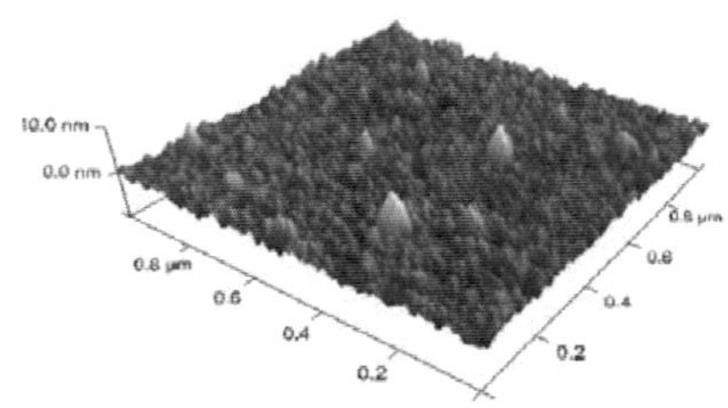

Figure 7. AFM picture of MoO_x thin film,deposited with solution M2 annealed at 200°C in nitrogen (Rq-0.8 nm).

3.3. Morphological characterization

Figure 6 presents the AFM image of sample MoO_x thin film, deposited with solution 184 mM $MoO_2(AcAc)_2$ in ethanol and annealed at 300°C in nitrogen. The estimated (average) grain size is about 200 nm x 40 nm and the root mean square (RMS) value of the surface roughness is about 5 nm. Deposition from diluted solutions leads to formation of flat and more uniform films. Figure 7 presents the AFM image of a MoO_x thin film, deposited from a diluted solution of 92 mM $MoO_2(AcAc)_2$ in ethanol and annealed at 200°C in nitrogen. Significantly lower values of the grain size (about 130 nm x 20 nm) and RMS surface roughness (1.0 ± 0.3 nm) were obtained for annealing temperatures in the range 100°C ÷ 250°C.

Table 1. Sheet resistance of MoO_x thin films, deposited with solutions of 184 mM(M1) and 92 mM (M2) $MoO_2(AcAc)_2$ in ethanol and annealed in nitrogen.

M1 Temp, °C	sheet resistance, Ω/□.	Thickness, nm	M2 Temp, °C	sheet resistance, Ω/□
50	265	56	50	
100	260	54	100	
200	380	58	200	260
250	277	58	250	321
300	258	34	300	n.a.

3.4. Ellipsometry and electrical properties
Spectral ellipsometric data (cf. figure 8) were used to determine the thickness of MoO_x films, deposited with solution M1. The values are summarized in Table 1 together with the data for estimated sheet resistances of the related layers.

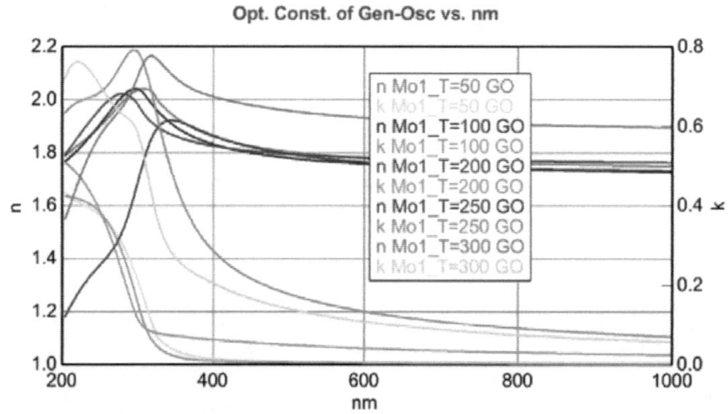

Figure 8. Dispersion of Refractive index and Extinction coefficient of MoO_x thin films, deposited with solution **M1** and annealed at different temperatures in nitrogen.

The sample deposited with concentrated stock solution and annealed at 300°C has remarkably lower thickness. This is probably due to the total desolvation and recrystalization of the MoO_x films. All samples show p-type conductivity and sheet resistance in the range of 260 – 380 Ω/□. Taking into account, that the resistivity $\rho = R_{sheet}$ x *film thickness* and using the data from table 1, we obtain values in the range of 10^{-5} Ω.cm, i.e. thin films of MoO_x could perform as excellent hole-conducting layers.

The spectral distribution of the refractive index and the extinction coefficient for samples deposited with solution 184 mM $MoO_2(AcAc)_2$ in ethanol and annealed at related temperatures in nitrogen is presented in figure 8. Estimated region for beginning of the strong absorption from curves of extinction is before 400 nm which coincides with the data for optical absorbance in figure 4.

Conclusions

The data obtained from DSC analysis of the active compound molybdenum (VI) oxide bis (2,4 – pentanedionate) and its related solutions with absolute ethanol were used to establish the regime for thermal treatments to form MoO_x thin films. Raman scattering have shown gradual desolvation and recrystalization of the layers and formation of definite orthorhombic MoO_3 at temperatures far above 300°C. Elaborated optical investigations revealed high transparency over 90 % with negligible fall near 770 nm (1.6 eV) for temperatures of annealing between 200 – 250°C and a recovery at higher temperatures. Estimated band gap of disordered MoO_x films was about 3.1-3.3 eV. The AFM data have shown decrease of grain sizes of thin MoO_x films with increasing of the annealing temperature and RMS surface roughness about1.0 ± 0.3 nm (as in [13]). The electrical measurements indicated definite p-type conductivity and sheet resistance of the respective thin films in the range 260 – 380 Ω/□. The obtained results show that the properties of solution processed MoO_x thin films are appropriate for application in organic solar cells.

Acknowledgements

This work was supported by Bulgarian National Science Foundation under contract TO2 – 12/ 12.12.2014. Authors express gratitude to Dr I.Bineva for AFM and Dr P.Terzijska for ellipsometric measurements in association with INERA.

Journal of Physics: Conference Series **764** (2016) 012011 doi:10.1088/1742-6596/764/1/012011

References

[1] Yu Wei Su, Shang Che Lan and Kung Hwa Wei. Organic photovoltaics, 15 *Materials today* **12** 554

[2] Kim H-S, Lee C-R, Im J-H, Lee K-B, Moehl T, Marchioro A, Moon S-J, Humphry-Baker R, Yum J-H, Moser JE, Gratzel M and Park N-G 2012 *Sci. Rep.* **2** 591

[3] Jacobsson T J, Correa-Baena J-P, Pazoki M, Saliba M, Schenk K, Gratzel M and Hagfeldt A, 2016 *Energy & Environ. Sci.* **9** 1706

[4] Yang W S, Noh J H, Jeon N J, Kim Y C, Ryu S, Seo J and Seok S Il, 2015 *Science* **348** 6240 1234

[5] Jeng J-Y, Chiang Y-F, Lee M-H, Peng S-R, Guo T-F, Chen P and Wen T-C 2013 *Adv.Mater.* **25** 3727

[6] Docampo P, Ball JM, Darwich M, Eperon GE and Snaith KJ 2013 *Nat.Commun.* DOI: 10.1038/ncomms3761

[7] Yeo JS, Kang R, Lee S, Jeon YJ, Myoung N, Lee CL, Kim DY, Yun JM, Seo YH, Kim SS and Na SI, 2015 *Nano Energy* **12** 96

[8] Kim JH, Liang PW, Williams ST, Cho N, Chueh CC, Glaz MS, Ginger DS and Ken AKY, 2015 *Adv.Mater.* **27** 695

[9] Hu L, Wang W, Liu H, Peng J, Cao H, Shao G, Xie Z, Ma W and Tang J 2015 *J.Mater.Chem. A* **3** 515

[10] Zilberberg K, Gharbi H, Behrendth, Trost A S and Riedl T 2012 *ACS Appl.Mater.Interfaces* **4** 1164

[11] Kirchartz T, Pieters BE, Kirkpatrick J, Rau U and Nelson J 2011 *Phys.Rev. B* **83** 115209

[12] MacKenzie RCI, Shuttle CG, Dibb GF, Treat N, Hauff E von, Robb MJ, Hawker CJ, Chabinyc ML and Nelson J 2013 *J.Phys.Chem.C* **117** 12407

[13] Jasieniak J, Seifter J, Mates JjoT and Heeger AJ 2012, *Adv.Funct.Mater.* **22** 2594

[14] Tremolet de Villers BJ, MacKenzie RCI, Jasieniak J, Treat ND and Chabinyc ML 2014 *Adv.Energy Mater* **4** 1301290

[15] Uzarevic K, Rubcic M, Radic M, Puskaric A and Cindric M 2011 *Cryst.Eng.Comm* **13** 4314

[16] Seguin L, Figlarz M Cavagnat R and Lassègues J-C 1995 *Spectrochim Acta A* **51** 1323

[17] Jasieniak J J, Seifter J, Jo J, Mates T and Heeger A J 2012 *Adv. Funct. Mate*r. **22** 2594

[18] Werner J, Geissbuhler J, Dabirian A, Nikolay S, Morales-Masis M, De Wolf S, Niesen B, Balif C, *ACS Appl.Mater.&Inter.* DOI : 10.1021/acsami.6b04425

[19] Lefteriotis G, Papaefthimiou S, Yanoulis P and Siokou A 2001 *Thin Solid Films* **384** 298

Journal of Physics: Conference Series **764** (2016) 012012 doi:10.1088/1742-6596/764/1/012012

How to avoid concentration unsteadiness with a CVD precursor delivery system employing bubbler?

G Peev and D Peshev

University of Chemical Technology and Metallurgy, Department of Chemical Engineering, 8 Kl. Ohridsky Blvd., 1756 Sofia, Bulgaria

E-mail: georgiapeev@gmail.com, d.peshev@uctm.edu

Abstract. An approach of great certainty is suggested as to avoid concentration unsteadiness with a CVD delivery system employing bubbler. Assuming that the carrier gas has to be saturated of precursor vapour during the entire CVD experiment a set of equations has been selected from the literature for calculating the height of liquid layer in the bubbler which is necessary to this aim. The selection is made on the prerequisites that gas bubbles are spherical, diffusion of precursor vapour in them is molecular and it takes place during their rising only, which has a velocity equal to the terminal bubble velocity of rising. The errors at levels lower than the one for saturation are discussed. Heights of $Al(CH_3)_3$ calculated using the approach are compared to those at which the bubbler controlling LED indicator of Dow Electronic Materials lights green, amber or red.

1. Introduction

The precursor delivery system is an important integral part of any CVD set-up. Very often this system utilizes bubbler wherethrough the carrier gas is passing and uptakes precursor vapour. In the case of multicomponent coating each component is usually delivered from a separate bubbler. The constant precursor concentration in the gas mixture entering the CVD reactor is unavoidable condition for correct research and production of high quality, therefore the precursor concentration in the carrier gas exiting the bubbler should be kept constant. The gas saturation has been approved as a basic principle for this constancy [1]. This requires maintaining a sufficiently high level of the liquid precursor in the bubbler. For this purpose, methods for control of the level [2] or to keep it constant [3] have been developed. The system with a self-metering reservoir is very complicated and in many cases – inapplicable [3]. Some companies offer various precursors in bubblers with visual liquid multi-level indicator for trademark (TM) vaporization delivery system and guarantee maximum source consumption [4]. The number of new liquid precursors however is increasing. The cost of the TM delivery systems and bubblers are very high and some researchers have to use in house made CVD set-up and bubbler. In such cases it is necessary to predict the precursor level at which the carrier gas will exit the vessel fully saturated during the entire deposition process.

The aim of the present work is to propose an approach which will provide such prediction with high certainty. This approach can be used in many other cases where gas-vapour mixture with constant concentration has to be prepared by bubbling. Typical examples are preparation of such mixtures for research in adsorption and catalysis as well as humidifying of gases.

Journal of Physics: Conference Series **764** (2016) 012012 doi:10.1088/1742-6596/764/1/012012

2. Theoretical analysis

The carrier gas in the bubbler is in contact with the precursor from the bubbles formation until their break-up on the liquid surface. At some conditions the breaking results in formation of small drops which may evaporate. Neglecting this vaporization and the mass transfer during the bubble formation and residence time in the foam layer is the first point of the approach.

The mass transfer from the liquid into the rising bubbles is a diffusion with a constant concentration at the interface, C_i, which is determined by the vapour pressure of the liquid at its temperature. Therefore the rate of mass transfer will depend on the diffusion resistance in the gas phase only. The latter depends on the bubble shape and hydrodynamic conditions in it. They are both influenced by the bubble size. The very small bubbles (d □ 1 mm) are spherical and the gas phase in them is immobile [5-7]. They behave like rigid spheres and the vapour penetrates in them by molecular diffusion. The differential equation for such unsteady-state diffusion with constant concentration at the interface has analytical solution valid when $R \to 0$ [8]

$$\frac{C-C_1}{C_i-C_1}=1+2\sum_{n=1}^{\infty}(-1)^n \exp\left(-Dn^2\pi^2 t/R^2\right) \tag{1}$$

According to the simplifications already made, the initial concentration, C_1, should be zero and the time of contact should be

$$t \geq H/w \tag{2}$$

In relation (2) H is the liquid height over the place of bubbling and w - the bubble velocity. The small spheres rise vertically and for them the sign of equality is applicable. The path of bigger bubbles is more complicated [5-7] but for them assuming the equality will increase the certainty of prediction. Accounting for these considerations Eq. (1) may be transformed:

$$\Delta=1-\frac{C}{C_i}=-2\sum_{n=1}^{\infty}(-1)^n \exp\left(-Dn^2\pi^2 H/wR^2\right) \tag{3}$$

Equation (3) has been numerically solved up to the nearly complete sphere saturation, i.e. for $C \in [0, C_i)$, at the sphere centre, and the results obtained are tabulated [7]. They are given in table 1 together with some values obtained from figure 6.1 in [8].

Table 1 Numerical results according to equation (3).

Dt/R^2	0	0.032	0.080	0.100	0.160	0.240	0.30	0.320	0.40	0.800
Δ	1	0.9975	0.8276	0.7071	0.4087	0.1871	0.104	0.0850	0.040	0.0007

It can be seen that at $\left(Dt/R^2\right)=0.800$, the concentration in the bubble centre will be $C=0.9993C_i$. At 0.4 C will be 96% of C_i, but the average concentration will be much nearer to C_i. The exact results can be found from the next equation [8]:

$$\frac{M_t}{M_\infty}=\frac{C_{av}}{C_i}=1-\frac{6}{\pi^2}\sum_{n=1}^{\infty}\frac{1}{n^2}\exp\left(-Dn^2\pi^2 t/R^2\right), \tag{4}$$

or by means of table 2 [7].

Table 2 Variation of dimensionless concentration, Δ', with Dt/R^2.

Dt/R^2	0	0.005	0.01	0.02	0.04	0.06	0.1	0.2	0.3	0.5	0.8
Δ'	1	0.774	0.690	0.579	0.440	0.352	0.226	0.084	0.031	0.0043	0.0002

Journal of Physics: Conference Series **764** (2016) 012012 doi:10.1088/1742-6596/764/1/012012

In table 2 Δ' is $(C_i - C_{av})/(C_i - C_1)$. At $C_1 = 0$ and $Dt/R^2 = 0.8$, $C_{av} = 0.9998C_i$. If the value of 0.8 is agreed as a condition for saturation of the bubbles, the necessary height to this aim, H, can be found from the relation

$$H = \frac{0.8wR^2}{D} \tag{5}$$

In the particular case of substance A diffusion into the gas B, which takes place in the process of saturation, the diffusivity coefficient, D, can be found from the relation of Fuller et al. [9]

$$D_{AB} = \frac{10^{-3}T^{1.75}\left(1/m_A + 1/m_B\right)^{1/2}}{P\left[\left(\Sigma\upsilon_A\right)^{1/3} + \left(\Sigma\upsilon_B\right)^{1/3}\right]^2} \tag{6}$$

or the relation of Maxwell/Gilliland [10]

$$D_{AB} = \frac{4.3\times10^{-3}T^{3/2}\left(1/m_A + 1/m_B\right)^{1/2}}{P\left[\left(\Sigma\upsilon_A\right)^{1/3} + \left(\Sigma\upsilon_B\right)^{1/3}\right]^2} \tag{7}$$

Experimental data are also available for some binary mixtures [9,10].

The rising velocity of a bubble, w, depends on its volume. The increasing of the volume results in a change of the bubble shape from spherical through ellipsoidal to semispherical and flat [5-7]. For a given gas-liquid system the shape can be controlled by the diameter of the bubbler holes, d_0, and the gas flow rate, V_0. If the latter is smaller than a critical value, V_{0c}

$$V_{0c} = \frac{\pi d^2}{6}w \tag{8}$$

and the liquid is wetting the bubbler material

$$d = \left(\frac{6d_0\sigma}{g\left(\rho_L - \rho_G\right)}\right)^{\frac{1}{3}} \tag{9}$$

Selecting d_0 one can obtain spherical bubble for which eqs. (3), (4) and Tables 1, 2 are applicable. As a first iteration $d \le 1 \div 1.5$ mm is a required condition [5-7].

The chosen value of d allows the calculation of rising velocity, w. In the case of bubble however there is no well-established opinion which equation is the most correct [5-7]. The rigid sphere behaviour found in several investigations has been attributed to the presence of surface active agents in the liquids used [5] and the Rybczynski-Hadamard relation for circulating drop adopted for bubble has been recommended at $\mathrm{Re} < 1$ ($gR^3/3\nu^2 < 1$) [5].

$$w = \frac{1}{3}\frac{\left(\rho_L - \rho_G\right)gR^2}{\mu} \tag{10}$$

The latter predicts higher velocity than Stokes relation for rigid sphere

$$w = \frac{2}{9}\frac{\left(\rho_L - \rho_G\right)gR^2}{\mu} \tag{11}$$

and thus increases the certainty of saturation. Regarding eq. (8) however eq. (11) has to be used and lower value of V_{0c} will be obtained. If the carrier gas flow rate of this value will be insufficient for the CVD, the precursor temperature, number of bubbler holes or both can be increased as to deliver the required precursor quantity.

The requirement $gR^3/3\nu^2 < 1$ together with eq. (9) results in very small bubbler holes corresponding to sintered materials. Cappilary tubes may produce bubble with diameter in order of

10^{-3} m, which may deviate from spherical but the equations and tables already discussed can be also applied. This is justified by the smaller path of penetration as well as by the increased circulations or oblate-prolate oscillations of the bubble which create convective diffusion.

For bubbles of such dimensions the value of Reynolds number is usually above 400 and the following relation can be used for calculating the velocity [5]:

$$w = 6.603\sqrt{R} \tag{12}$$

It is important to note that eqs. (11) and (12) predict the terminal rising velocity after a short period of acceleration, therefore the real bubble residence time will be somewhat longer. The bubbles successively obtained are not exactly alike.

The diameter and velocity of the bubbles can be determined experimentally as well. Techniques to this aim have been well developed.

It should be pointed out that the concentration unsteadiness is not very sensitive to the exact calculation of liquid height in the bubbler. Data in Table 2 show that if value of 0.2 is chosen instead of 0.8 for calculating the height above which the liquid level should be maintained, the maximum change of the average precursor concentration will be of 8.4 % below the saturation. At 0.3 it will be 3.1 %.

3. Some illustrative numerical results

To evaluate the suggested approach we shall use the available information about DOW bubblers with liquid multi-level indicator [4]. LED lamps of three colors light when the level is: green – above 20 %, amber – between 20 % and 10 % and red – below 10 % of the initial precursor height respectively. For TM bubbler with height of 108 mm, diameter of 51 mm and liquid volume of 150 ml, the initial level is 80 mm and the levels at which the respective lamp will be lit-up are >16, 16-8 and <8 mm. For TM bubbler with precursor volume of 3000 ml, the respective dimensions are 254, 127 and 240 mm. The levels for the respective color are >48, 48-24 and <24 mm. In Table 3 the minimum levels required for saturation of nitrogen with $Al(CH_3)_3$ (TMA) vapor, calculated from the respective equations are shown for two often used capillary diameters.

The TMA diffusivity coefficient is found from Eq. (7). The other characteristics are as follows: surface tension – 0.0261 N m^{-1}, viscosity – 0.94×10^{-3} Pa s and density – 750 kg m^{-3}. All data are at temperature of 28 ^{o}C.

The results of Table 3 seem very realistic. Experiments for a more precise evaluation of the approach will be aim of a future work.

Table 3 Illustrative calculations for the levels of $Al(CH_3)_3$ allowing comparison with DOW proposals.

№	capillary diameter, d_0, mm	bubble diameter, d, mm Eq. (9)	rising velocity, w, m s^{-1} Eq. (12)	Reynolds number Re	required level, H, mm Eq. (5)	level at deviation of 3% from saturation, mm
1	0.5	2.21	0.220	387	>23.6	8.8
2	1	2.79	0.246	547	>42	15.7

4. Conclusions

The theoretical analysis of the problem has shown that as to avoid a concentration unsteadiness the precursor level in a bubbler must be kept above a certain critical value during the entire deposition process. This value depends on many factors (diameter of the bubbler holes, precursor density, viscosity, surface tension and diffusivity in the carrier gas, bubble shape and dimensions, gas flow rate, temperature, etc.) which have to be accounted for with each particular case. An approach for calculating this value is suggested. It reveals a moderate sensitivity of the unsteadiness from the changes of this level, but they should be considered regarding the precision of CVD experiments.

Notations

C	current local precursor concentration (kg m^{-3})
C_1	initial precursor concentration in bubble (kg m^{-3})
C_{av}	mean current precursor concentration (kg m^{-3})
C_i	concentration of saturation (kg m^{-3})
D	diffusivity coefficient of the precursor vapour in the carrier gas (m^2 s^{-1})
d	bubble diameter (m)
d_0	diameter of the bubbler holes (m)
g	Earth acceleration (m^2 s^{-1})
H	height of liquid above the place of bubble formation (m)
M_A	molecular mass of the carrier gas (kg mol^{-1})
M_B	molecular mass of the precursor (kg mol^{-1})
M_t	precursor mass in the bubble at time t (kg)
M_∞	precursor mass in the bubble at saturation (kg)
P	pressure (atm)
R	radius of the bubble (m)
Re	Reynolds number, $\frac{wd\rho_L}{\mu}$ (-)
T	temperature (K)
t	residence time of the bubble in the bubbler liquid (s)
V_{oc}	critical carrier gas flow rate (m^3 s^{-1})
V_o	carrier gas flow rate (m^3 s^{-1})
v	kinematic viscosity (m^2 s^{-1})
w	terminal rising velocity of the bubble (m s^{-1})
μ	precursor viscosity (Pa s)
ρ_G	carrier gas density (kg m^{-3})
ρ_L	precursor density (kg m^{-3})
σ	surface tension precursor/carrier gas (N m^{-1})
υ_A	molecular volume of the carrier gas
υ_B	atomic volume of the precursor elements

References

[1] Jones AC and Hitchman ML 2009 *Chemical Vapor Deposition: Precursors, Processes and Applications* (Cambridge: The Royal Society of Chemistry, RSC Publishing) p 81
[2] Bour D, Yang Zh and Chua Ch 2008 Simple technique for measuring the filled volume of liquid or solid CVD precursor chemicals in bubblers *J. Crys. Growth* **310** 2673-77
[3] O'Brien P, Pickett NL and Otway DJ 2002 Developments in CVD delivery systems: a chemist's perspective on the chemical and physical interaction between precursors *Chem. Vap. Depos.* **8** 237-49
[4] Chemical Vapor Deposition Sources, DOW Electronic Materials, 33795, Rev. 0, April 2013, p 3.
[5] Levich V G 1962 *Physicochemical hydrodynamics* (Englewood Cliffs, NJ: Prentice Hall)

Journal of Physics: Conference Series **764** (2016) 012012 doi:10.1088/1742-6596/764/1/012012

Chapter 8
[6] Ramm VM 1966 *Absorbcija gazov* (Moskva: Himija) pp 511-516 (in Russian).
[7] Sherwood TK, Pigford RL and Wilke CR 1982 *Mass Transfer* (Moskva: Himija) pp 91-92, 265-267 (Russian translation)
[8] Crank J 1967 *The mathematics of diffusion* (London: Oxford University Press) pp 84-87.
[9] Reid RC, Prausnitz JM and Sherwood TK 1982 *The properties of gases and liquids 3d edition* (Leningrad: Himija) pp 475-480 (Russian translation)
[10] Pavlov KF, Romankov PG and Noskov AA 1976 *Primeri i zadachi po kursu processov i apparatov himicheskoj tehnologii* (Leningrad: Himija) pp 263-264, 515 (in Russian).

Journal of Physics: Conference Series **764** (2016) 012013 doi:10.1088/1742-6596/764/1/012013

Synthesis and characterization of thin amorphous carbon films doped with nitrogen on (001) Si substrates

I Balchev[1], Kr Tzvetkova[1], S Kolev[1], P Terziiska[2], A Szekeres[2], I Miloushev[2], T Tenev[2], K Antonova[2], R Peyeva[2], T Ivanova[3], I Avramova[4], M Tzvetkov[5], G Avdreev[5], E Valcheva[6], T Milenov[1] and S Tinchev[1]

[1]Institute of Electronics- Bulgarian Academy of Sciences, 72 Tzarigradsko Shausee Blvd., 1784 Sofia, Bulgaria
[2]Institute of Solid State Physics- Bulgarian Academy of Sciences, 72 Tzarigradsko Shausee Blvd., 1784 Sofia, Bulgaria
[3]Central Laboratory of Solar Energy and New Energy Sources- Bulgarian Academy of Sciences, 1784 Sofia, Bulgaria
[4]Institute of General and Inorganic Chemistry, Bulgarian Academy of Sciences, Acad. G. Bonchev Str., bl. 11, 1113 Sofia, Bulgaria
[5]"R. Kaishev" Institute of Physical Chemistry, Bulgarian Academy of Sciences, Acad. G. Bonchev Str., bl. 11, 1113 Sofia, Bulgaria
[6]Faculty of Physics, Sofia University, 5 James Bourchier Blvd., 1164 Sofia, Bulgaria

E-mail: ivobalchev@gmail.com

Abstract. We synthesized undoped as well as up to 6 at.% nitrogen (N) doped thin amorphous carbon (α-C) films by plasma- enhanced chemical vapor deposition (PECVD) method. The source of carbon/carbon containing radicals was benzene (C_6H_6) in Ar gas- mixture. The obtained thin films were studied by optical microscopy, X-ray powder diffraction, UV-VIS-NIR Spectral Ellipsometry, IR and Raman spectroscopic studies as well as by X-ray photoelectron spectroscopies (XPS). We established by XPS that the deposited layers consist of a mix of sp^2 and sp^3 hybridized carbon. The films are amorphous as it was shown by the measured XRD patterns. The ellipsometric measurements enabled calculation of transition energies and the complex results showed that films with thickness of 15- 120 nm and different properties can be obtained by this technique.

1. Introduction

Over the last decades carbon-based materials and α-C (sp^2- dominated amorphous carbon films) and ta-C(sp^3 dominated diamond-like carbon films) especially are extensively investigated for use in electronic applications. The plasma enhanced chemical vapor deposition (PECVD) method is the most widely used due to its high reliability, relatively low cost etc. It is established thatthe optical bandgap $E_g$of the deposited films can be easily varied with the plasma voltage [1].Additionally, the charge transport properties can be modified by doping.Nitrogen (N) looks perfect candidate as it has very similar size as carbon and belongs to III group of the Periodic system. Moreover, the nitrogen level in forbidden gap of ta-C is shallow [2].Different carbon containing precursors are investigated but methane is the preferred one due its relatively high ionization potential and low deposition rate [3,4]. Conversely, the application of benzene as a precursor is usually avoided due to its low ionization potentialand high deposition rate of about 2 nm/s or higher [5-7]. Moreover, the charge transport in this case is most probablygoverned by hopping conduction in the valence band tail[5].

We aimed in the deposition of N-dopedhydrogenated carbonfilms with different sp^2/sp^3ratio by PECVDat low deposition rates for application in gas-detectors.Here we present some preliminary results of characterization of several representative specimens of hydrogenated ta-C and α-C doped with 0.1- 1 at.% N.

2. Experimental

Hydrogenated ta-C (ta-C:H) and α-C (α-C:H)thin films were deposited in a DC plasma CVD reactor. We used benzene as a hydrocarbon precursor and an argon residual pressure. More information about the used process and equipment can be found in [1]. The Si substrates were placed on cathode while the metal vacuum chamber acted as anode. Benzene vapourwas obtained in a thermostat at about 25^0C. Before the film deposition the substrates were cleaned in acetone, ethanol and deionized water in ultrasonic bath followed by a 10 min Ar^+ ion cleaning at 2 kV. The deposition processes were performed in DC plasma (U= (1400- 1800)V and I= 75mA) at Ar residual pressure of 4.10^{-2}Torr.The nitrogen (99,995 % pure N_2) gas is introduced in the reaction chamber through a needle valve. The specimens E_S1 and E_N3 are deposited at about 0.9 nm/s while E_4 at 2 nm/s, respectively.

Thin films were characterized by ellipsometry, XRD, Raman, IR and XP spectroscopy. The ellipsometry measurements were performed using J.A. Woollam Co., Inc. M2000D rotating compensator spectroscopic ellipsometer with a CCD spectrometer with wavelength range from 193 to 1000 nm. Experimental data for psi and delta were acquired at angles of incidence of 55, 60, 65, 70 and 75 degrees and were modelled using the CompleteEASEWoollam Co., Inc. software. The Raman measurements were carried out in a micro-Raman spectrometer HORIBA Jobin Yvon Labram HR 800 Visible with a He-Ne (633 nm) laser. The laser beam with 0.5 mW power was focused on a spot of about 1 μm in diameter on the studied surfaces, the spectral resolution being 0.5 cm^{-1} or better.The X-ray powder diffraction patternstaken from {001} plane for phaseidentification were recorded in the angle range $2\theta = 24^0– 61^0$(about <011> at $\varphi=0^0$ as well as at $\varphi=90^0$) on a Philips PW 1050 diffractometer, equipped witha Cu Kα tube at temperature of 21^0C. A step-scan mode at steps of $0.05^0(2\theta)$ with a counting time of 1 s/step was used for theXRD data collection. The X-ray Photoelectron spectra were obtained using non-monochromatized Al K_α (1486.6 eV) radiation in a VG ESCALAB MK II electron spectrometer under base pressure of $1x10^{-8}$ Pa. The spectrometer resolution was calculated from the $Ag3d_{5/2}$ line with the analyzer transmission energy of 20 eV. The full width at half maximum (FWHM) of this line is 1eV. The spectrometer was calibrated against the $Au4f_{7/2}$ line (84.0 eV) and the samples' charging was estimated from C1s (285 eV) spectra from natural hydrocarbon contaminations on the surface. The accuracy of the BE measured was 0.2 eV. The photoelectron spectra of C1s, O1s, Fe2p, Ni2p as well as CKVV Auger line of carbon films deposited on different substrates were recorded and corrected by subtracting a Shirley-type background and quantified using the peak area and Scofield's photoionization cross-sections.The IR spectra were recorded using a Bruker VERTEX 70 System. The signal was averaged over 64 scans in the spectral range 600-6000cm^{-1}. The spatial resolution was 4 cm^{-1}.

3. Results and discussion

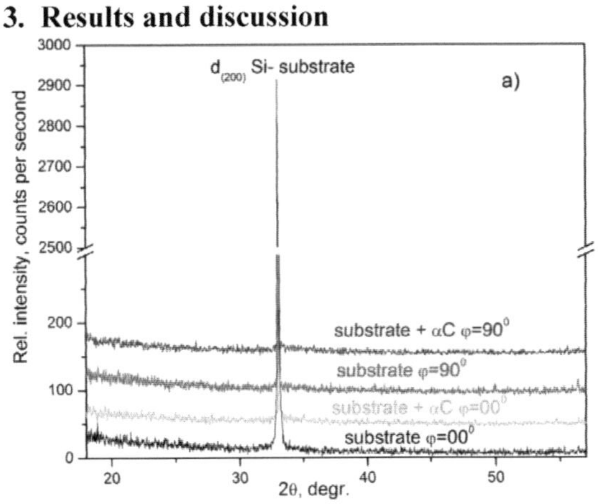

Figure 1 a) XRD pattern of the Si-substrate(black and red traces) as well as the E_4 (greenand blue traces) specimens.

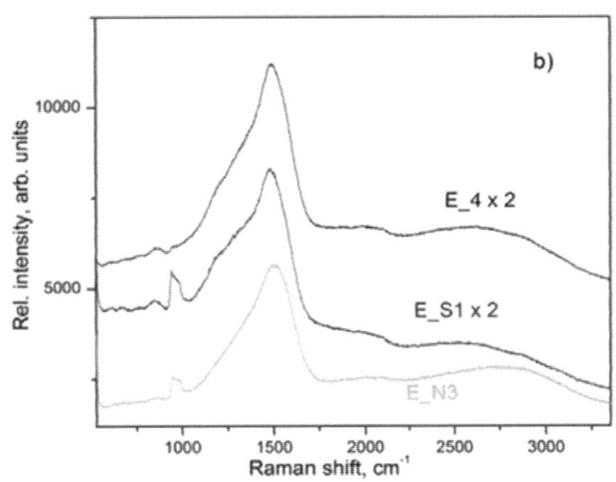

Figure 1 b) Raman spectra of E_4, E_S1 and E_N3 specimens.

The XRD patterns of E_4 specimen (before and after the deposition process) show absence of long range ordering in the carbon layer and only Si (002) reflection appears in the figure 1a). The XRD patterns of E_N3 and E_S1 specimens are not shown here as theyare very similar to those of E_4. The Raman spectrum of the specimens E_4, E_N3 and E_S1 (figure 1b) completely coincide with those of hydrogenated amorphous carbon: it is dominated by a broad feature situated at about 1450 cm^{-1}[8].

The IR absorption spectra (figure 2) consist of several weak absorption bands indicated by consequent numbers from 1 to 6, which should be ascribed as follows: 1 and 2- are due to mono-substituted benzene (CH_3-C_6H_5)[9];3 and 4- most probably are due to sp^3 C-H stretching in aliphatic(-CH_2-..-CH_2-) chains [10]; 5- is the asymmetric vibrationυ_3 of $[SiO_4]^{2-}$ group [11] and 6- alkyl-substituted ether (RA-O-RA) species[9].

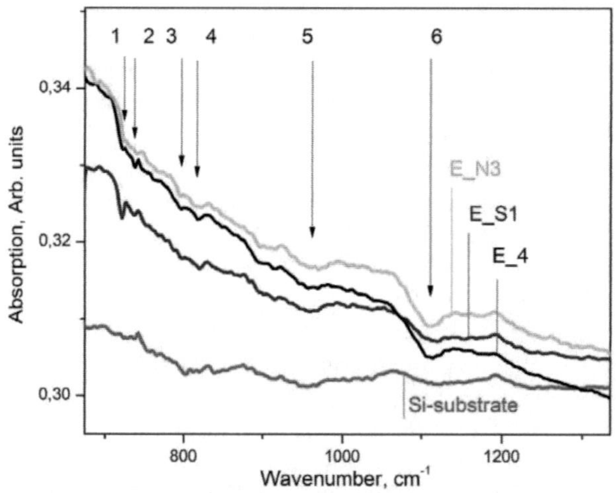

Figure 2.IR absorption spectra of the Si- substrate, E_4, E_N3 and E_S1 specimens.

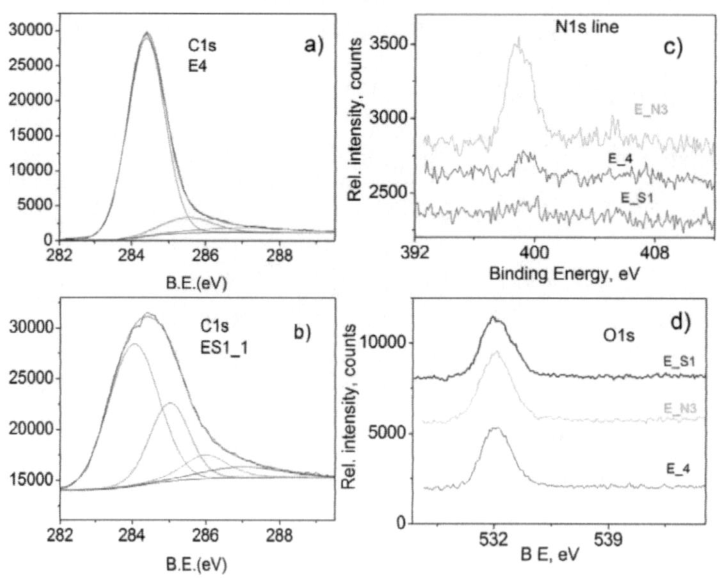

Figure 3. XPS results of E_4 and E_S1 specimens: **a)** and **b)** panelsrepresentC1s lines;**c)**shows N1s lines and **d)** shows O1s lines.

The C1s spectrum of amorphous carbon films (figures 3a and b) is dominated by a single peak at 284.2 eV (corresponding to sp^2 hybridized carbon) and a second peak at the higher binding energy side in the spectra having binding energy at around 284.9 eV (corresponding to sp^3 hybridized carbon) [12,13].The deconvolutions of 1s carbon spectra gave us possibility to calculate the sp^3/sp^2 ratio and to extract additional information about the carbon to carbon bonds as well as the carbon to oxygen bonds.N1s lines (figure. 3c) pointed to a different concentrations and forms of doping nitrogen: the specimens E_4 and E_S1 contain smaller quantities of nitrogen and the N1s line appears at about 400 eV therefore, it could be ascribed to pyrroleic nitrogen (C_4H_4NH). We associate the last small peak at the higher binding energy side in the N1s spectrum with existence of NO_2 (adsorbed species). The N1s line of E_N3 specimen is very complex but it seems that it contains pyridone nitrogen($C_5H_4NH(O)$) predominantly. The O1s lines (figure 3d) indicated that the oxygen in the carbon layer of E_4 specimen is below 3 at. % as we observed oxygen as a part of SiO_2 interlayer. The E_S1 and E_N3 carbon layers contain about 3.5 and 2 at. % oxygen, respectively.

For the samples (E_4, E_S1 and E_N3) we detected silicon in none fully oxidation states (the XPS are not shown here). At the same time for the samples ES1 and E_N3 we observed the presence of silicon in SiO_2. The presence or absence of silicon in the investigated by us highly tetrahedral amorphous carbon films is due to a different thickness of the deposited films as well as due to the useddeposition method. The XPS results are summarized in table 1.

Table 1. Summarized results for the layers configuration and thickness as well as for their phase composition obtained by XPS and ellipsometry studies.

Experiment	**Fraction sp^3/sp^2 C**	**C,at%**	**O,at%**	**N,at%**	**Layer configuration Thickness, nm**	
					SiO_2	**ta-C:H**
E_4	0.1	93.56	5.95	0.42	320.9	138.4
E_N3	0.55	90.19	8.48	1.33	427.5	26.2
E_S1	0.42	93.65	6.03	0.23	2.89	114.2

The measured ellipsometry spectra were fit in two different ways depending on the properties of the films:

-samples E_N3 and E_4 (the ta-C:H andα-C:H layersare deposited on a300-400 nm thick thermal SiO_2 layer): a 3-layer model was used to represent the sample, consisting of a silicon substrate, a thermal silicon oxide layer (as a 1st layer), an α-C layer as a second layer and a roughness layer as a third layer. Each layer is determined by its thickness and dielectric constants. For the silicon substrate and the thermal silicon oxide layer we used the models Si_JAW and SiO2_JAW from the CompleteEASE data analysis software library. The top α-C layer was first represented as a b-spline layer with starting material Graphite_o_Palic_nk from CompleteEASE data analysis software library. The thickness of the thick thermal SiO_2 layer was determined by adjusting the number of the psi and delta oscillations and their amplitude was adjusted by varying the thickness of the alpha-C layer deposited on the top. This method is known as "interference enhancement method" and allows a unique determination of both the thickness and the optical constants of thin absorbing films on opaque substrates from VASE data [14]. The dielectric function of the alpha-C layer described by the b-spline model was then parametrised using the General oscillator(Gen-Osc) model including one Cody-Lorentz and one Gaussian oscillator. The Gen-Osc parametrization was used to reduce the numberof theunknown fit parameters as well as to ensure Kramers-Kroning consistent optical properties [15,16].

-sample E_S1 (the ta-C:H layer is deposited on a 2-4 nm thick native SiO_2): a 3-layer model was used to represent the sample, consisting of a silicon substrate, a silicon native oxide layer (as a 1st layer), an alpha-C layer as a second layer and a roughness layer as a third layer. For the silicon substrate and the native silicon oxide layer we used the models Si_JAWand NTVE_JAW from the CompleteEASE data analysis software library. The thickness of the silicon native oxide layer was determined on a bare substrate before the deposition of the top ta-C:H layer. The layer was then parametrised using b-spline model to fit the optical constants in the absorbing area. Finally, the dielectric function of the b-spline layer was parametrized using the General oscillator model including one Cody-Lorentz and one Gaussian oscillator.

The roughness layer for all samples is modeled by Bruggeman's EMA (Effective Medium Approximation) of 50% voids and 50% bulk material [16].

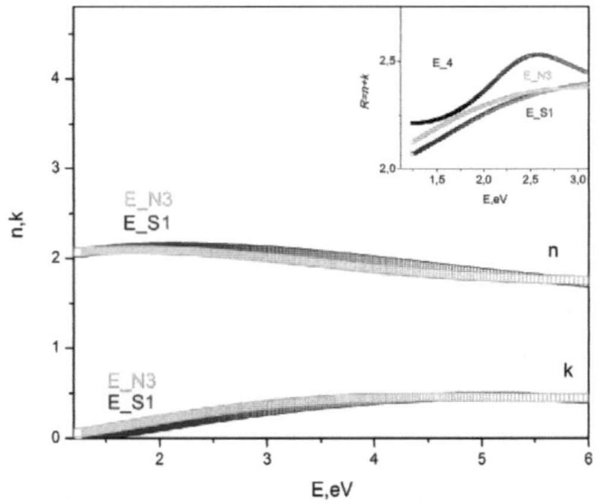

Figure 4. Optical parameters: R-refractive index, n- refraction coefficient and k-extinction coefficient obtained from the ellipsometry measurements.

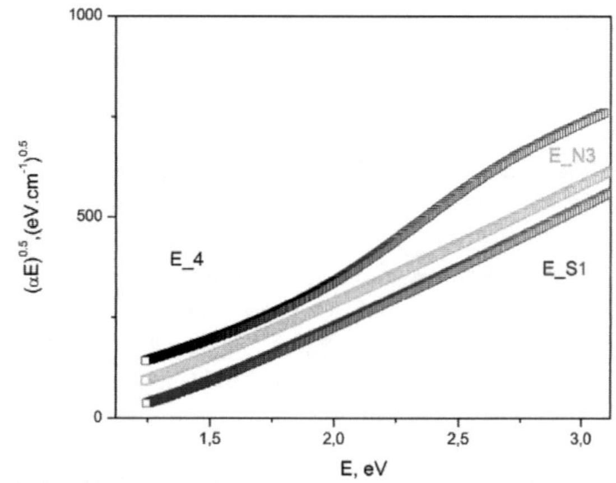

Figure 5. Calculated energy of indirect interbandtransitions (by Tauc relation: $\alpha E = A(E-E_g)^{0.5}$) [13] in E_4, E_S1 and E_N3.

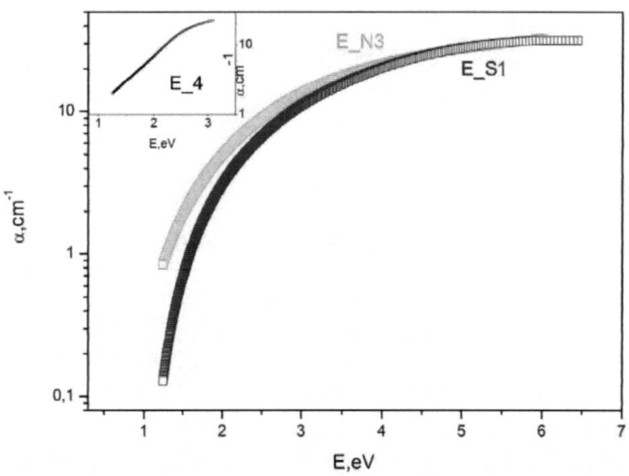

Figure 6. Energy of the indirectinterband transitions in E_S1 and S_N3 specimens and the defect Urbach-tail states (the inset) in E_4 specimen.

The spectral dependence of the basic optical parameters of the films (*R*- refractive index $R=n+k$, *n*-refraction coefficient and *k*- extinction coefficient) determined by that way are plot in figure 4. Furthermore, the absorption coefficient (α), the energy of indirect interbandtransition- E_g (shown in figure 5 and 6 as well as the excitation energy of the defect Urbach tail states- E_u (shown in the inset of figure 6) were calculated from these datasets by relations [17, 18, 19]:

$$\alpha = 4\pi k/\lambda \quad (1)$$

$$\alpha E = A(E-E_g)^{0.5} \quad (2)$$

$$\alpha(E)=\alpha_0 exp((E-E_g)/E_u) \quad (3)$$

Then, the energy of indirect interband transition in specimens E_N3 and E_S1 is about 1.07 and 1.27 eV, respectively- see figures 5 and 6. The energy of defect driven Urbach-tail states in specimen E_4 is about 0.93 eV- see the inset of figure 6.We did not observe tail states in specimens E_N3 and E_S2. This fact can be related to the low deposition rate and minimized disorder in the deposited films.It seems also that the nitrogen occupies different positions in the carbon framework depending on the dominating hybridization of carbon, thus resulting in different energy levels in the bandgap and charge transport properties of the doped films.

4. Conclusions

We deposited successfully nitrogen doped ta-C:H and αC:H layers with thicknesses (20-150 nm) and nitrogen concentrations from 0.2 to 1.5 at.%.The mole fraction of sp^2 to sp^3 hybridized carbon content could be varied by the DC- plasma voltage in the range 90/10 to 60/40 at. %.The layers are amorphous and contain predominantly C_6H_6; CH_3-C_6H_5 and sp^3 C aliphatic(-CH_2-..-CH_2-) species. The energy gap of indirect transitions in E_N3 and E_S1 layers is about 1.07 eV and 1.27 eV, respectively. The excitation energy of the defect Urbach tail states is about 0.93 eV in E_4 specimen. The lowered deposition rate from 2nm/s to below 1 nm/s decreases disorder and related tail states.

Acknowledgements:
The authors gratefully acknowledge the support from INERA "Research and Innovation Capacity Strengthening of ISSP BAS in Multifunctional Nanostructures" Project as well MPNS COST ACTION MP -1204.

References

[1] Tinchev S, Nikolova R, Dyulgerska J, Danev G and Babeva Tz 2005 *Solar Energy Materials Solar Cells* **86** 421
[2] Robertson J and Davis C 1995 *Diamond Rel. Mater.***4** 441
[3] Robertson J 2002 *Mater. Sci. Engineer. R* **37** 129
[4] Koidl P, Wild C, Dischler B, Wagner J, Ramsteiner M 1990 *Materials Science Forum* **52-53** 41
[5] Kalish R, Amir O, Brener R, Spits R A and Derry T E 1991 *Appl. Phys. A* **52** 48
[6] Han H X and Feldman B J 1988 *Solid State Commun.* **65** 921
[7] Kaufman J H, Metin S and Saperstein D D 1989 *Phys. Rev. B* **39** 13053
[8] Ferrari A and Robertson J 2004 *Phil. Trans. R. Soc. Lond. A* **362** 2477
[9] *Interpretation of Infrared Spectra, A Practical Approach by John Coates in Encyclopedia of Analytical Chemistry* (John Wiley & Sons Ltd, Chichester 2000) pp 10815-10837
[10] Chevalier J, Bergeron J and Dao L 1992 *Macromolecules* **25** 3325
[11] Nakamoto K 1989 *Infrared and Raman spectra in Inorganic and Coordination Compounds* (New York: John Willey & Sons.) IV 138
[12] Lascovich J, Giorgi R and Scaglione S 1991 *Appl. Surf. Sci.* **47** 17
[13] Díaz J, Paolicelli G, Ferrer S and Comin F 1996 *Phys. Rev. B* **54** 8064
[14] McGahanW, Johs B and Woollam J 1993 *Thin Solid Films* **234** 443
[15] Hilfiker J, Singh N, Tiwald T, Convey D, Smith S, Baker J and Tompkins H 2008 *Thin Solid Films* **516** 7979
[16] *CompleteEASE®* Software Manual (Lincoln: J.A.Woollam Co.) USA
[17] Dresselhaus M 2001 *Solid State Physics, p.II,Optical Properties of Solids*, p.36
[18] Tauc J, Grigorovici R and Vancu A 1966 *Phys. Status Solidi* **15** 627
[19] Urbach F 1953 *Phys. Rev.* **92** 1324

Journal of Physics: Conference Series **764** (2016) 012014 doi:10.1088/1742-6596/764/1/012014

Antimicrobial effect of TiO_2 doped with Ag and Cu on Escherichia coli and Pseudomonas putida

O Angelov[1], D Stoyanova[1,2] and I Ivanova[2]

[1]Central Laboratory of Solar Energy and New Energy Sources, Bulgarian Academy of Sciences, 72 Tzarigradsko Chaussee, 1784 Sofia, Bulgaria
[2]Department of "General and Industrial Microbiology", Faculty of Biology, Sofia University "St. Kl. Ohridski", 8 D. Tzankov, 1164 Sofia, Bulgaria

E-mail: oangelov@phys.bas.bg

Abstract. Antimicrobial effect of TiO_2 doped with Ag and Cu on Gram-negative bacteria *Escherichia coli* and *Pseudomonas putida* is studied. The thin films are deposited on glass substrates without heating during the deposition by r.f. magnetron co-sputtering of TiO_2 target and pieces of Ag and Cu. The studied films, thickness about 65 nm, were as deposited and annealed (520^0C, 4h, N_2+5%H_2, 4Pa). The as deposited thin films TiO_2:Ag:Cu have band gap energy of 3.56 eV little higher than the band gap of crystalline anatase TiO_2 which can be explained with the quantum effect of the granular structure of r.f. magnetron sputtered films. The annealed samples have band gap of 2.52 eV due to formation of donor levels from Ag and Cu atoms near the bottom of the conduction band. The toxic effect was determined through the classical Koch's method and the optical density measurements at λ=610 nm. The as deposited TiO_2:Ag:Cu thin films demonstrate stronger inhibition effect - bactericidal for *P. putida* and bacteriostatic for *E. coli* (up to the 6th hour) in comparison with the annealed samples. The both methods of study show the same trends of the bacterial growth independently of their different sensitivity which confirms the observed effect.

1. Introduction

TiO_2 is a widely used material for scientific study and application in different fields due to its high stability at chemical treatments and environment influence. Transition metals (W, V, Ag, Cu, etc.) or non-metals (S, C, N_2) doped TiO_2 has lower band gap than pure TiO_2 and can be used to increase the production of reactive oxygen species generated under visible light irradiation in solutions containing undesired pollutants [1, 2]. This modification of TiO_2 band gap doesn't affect its ultraviolet light activity and increases the photocatalytic activity at sunlight. TiO_2 can be successfully applied as antimicrobial agent for deactivation of different microorganisms in waste water, wound dressings, medicine, incubator walls for the neonatology, etc., [3]. Surface to volume ratio of the nanomaterials plays an important role in application of TiO_2 so the material is mainly in the form of nanoparticles which are usually dispersed in solution. However, this leads to segregation of the nanoparticles in dependence on their sizes, mass, temperature, pH and other environmental characteristics, which requires stabilization and homogenization of the nanoparticles [4, 5]. This invokes the necessity for synthesis and study of different structures of TiO_2 for optimization of its application as bactericide agent. Akhavan has studied the role of Ag-TiO_2/Ag/a-TiO_2 nanocomposite film prepared by sol-gel method for photodegradation of *Escherichia coli* [6]. He established OH^- bounds and H_2O contents on the film surface and at the interface between layers. These species stimulate the release of more ionic silver than Ag^0 atoms due to inter-diffusion of H_2O and Ag nanoparticles through pores of the TiO_2. The Ag^+ are responsible for the antibacterial effect of this nanocomposite thin film structure. The authors of [7] compared antimicrobial activity of TiO_2/Ag, TiO_2/Cu and co-deposited TiO_2-CuO with single layers. They established good antibacterial effect of Cu-TiO_2 films under irradiation through photo-Fenton type reaction (creation of free radicals via intracellular reactions of chemicals which are present in vivo). Thin films TiO_2 on surfaces in sterile zones, incubator walls, etc., could provide static antimicrobial cover, which is durable and active for a long time [8].

The aim of this article is to compare the antimicrobial effect of TiO_2 thin films doped with Ag and Cu on *Escherichia coli* and *Pseudomonas putida*.

2. Experimental

Thin films TiO_2 are deposited on glass substrates without deliberately heating during the deposition by r.f. magnetron co-sputtering (13.56 MHz) of TiO_2 target and placed together Ag and Cu pieces on the target surface in its maximum erosion zone. The total surface area of Ag (4 alike pieces) is 40 mm^2 and this one of Cu (4 alike pieces) is 80 mm^2. The r.f. power is 50 W and the sputtering atmosphere is Ar (0.8 Pa). The thickness of the films is about 65 nm and is measured with profilometer Taylor Hobson. The studied films were as deposited and annealed in N_2+5%H_2 atmosphere, 4 Pa, at 520^{0}C for 4 h. The optical properties of the films were analyzed through transmittance and reflectance measurements. The structure of the films were characterized by Raman spectroscopy.

The microorganisms in this study were supplied by the National Bank for Industrial Microorganisms and Cell Cultures (NBIMCC): *E. coli* 3548 NBIMCC (ATCC 10536) and *Pseudomonas putida* 1090 (ATCC 12633) by NBIMCC.

The thin film antimicrobial effect was studied by observation of bacterial growth in periodic culture. The toxic effect was determined through the classical Koch's method (plating on solid nutrient medium and counting of survived cells) and the optical density measurements (OD) at λ=610 nm. During the experiment through the Koch's method the control and samples were irradiated with a lamp Vivalux of 20 W at a distance of about 10 cm. The experiment for OD measurements was conducted in a six-well plastic plate. Series of two replicates for the controls with bacteria and pure glass without thin film, three replicates for tested thin films and bacteria and an empty control only with nutrient medium and a thin film were prepared. The inoculum quantity in all variants was constant. The optical density was measured every hour and in every 2 hours consecutive decimal dilutions were prepared for determination of the bacterial quantity of the samples and the controls. Each dilution is done in duplicate. The measurement of optical density was carried out with spectrophotometer Spekol 11 at λ=610 nm every hour up to 24 hours.

3. Results and discussion

Figure 1 displays the dependence of $(\alpha.E)^{0.5}$ (absorption coefficient α, multiplied by photon energy, E) on energy for as deposited and annealed thin films TiO_2:Ag:Cu. The absorption coefficient is calculated through transmittance and reflectance spectra in the energy range corresponding to the fundamental absorption [9]:

$$\alpha = (1/d)\ln[(1-R)^2/T] \qquad (1),$$

where α- absorption coefficient, d-film thickness, [cm], R-reflectance and T-transmittance.

The energy band gap is determined by the Tauc's equation [10] for indirect electron transitions typical for TiO_2. The band gap of the as deposited TiO_2:Ag:Cu thin films determined by the cross point between the tangent to the plot at the range near the absorption edge and the coordinate axis of energy, E, is about 3.56 eV, little higher than the band gap of anatase TiO_2 (~ 3.2 eV). This can be explained with the quantum size effect [11] due to granular structure of amorphous TiO_2 deposited by r.f. magnetron sputtering. At annealing the Ag and Cu atoms coalesce through diffusion inside the layer and create nanoparticles with different sizes and shapes because the interaction between the metal atoms is higher than between the individual atoms and TiO_2 layer. Raman spectra (not shown) demonstrate amorphous structure of the as deposited and annealed films. In the transmittance spectrum (not shown) of annealed samples there is a dip narrow band centered at about 400 nm. This band is ascribed to intensive plasmon resonance of Ag nanoparticles due to overlap between quadrupole and dipole resonance [12]. The position of this band depends on the size of nanoparticles and dielectric function of the TiO_2 layer. At wavelengths lower than 400 nm the transmission of the annealed samples is due to intensive quadrupole resonance of the Ag nanoparticles. The quadrupole resonance emits at about 365 nm stronger in forward direction, thus the signal from the sample is registered by the detector as transmission of the sample. At these conditions the denominator of the equation (1) increases the nominator is low and the logarithm is not defined with argument near to 0. This range corresponds to the discontinuity in the dependence of the absorption coefficient α on energy, E (3.68-3.9 eV). The band gap of the annealed TiO_2:Ag:Cu films is 2.52 eV (λ=492nm) due to formation of donor centers of the doping transition metals Ag and Cu atoms in the band gap near the bottom of the conduction band. This leads to visible light absorption of the TiO_2:Ag:Cu films which is potential for improvement of their UV and visible light photocatalytic activity.

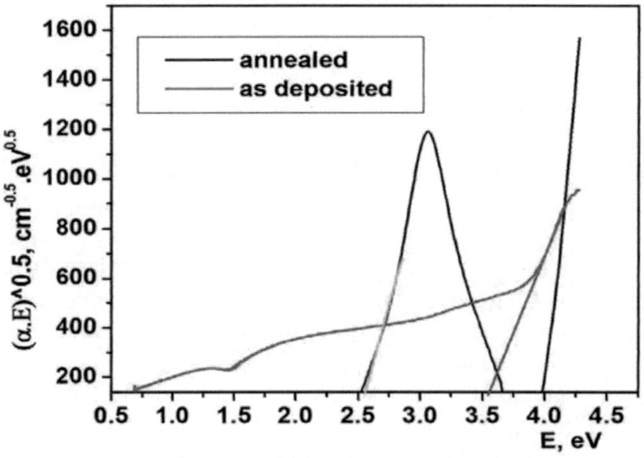

Figure 1. Dependence of $(\alpha.E)^{0.5}$ on photon energy, E, for TiO_2:Ag:Cu, as deposited – red line and annealed – black line.

Figure 2 demonstrates the bactericidal effect of as deposited thin films TiO_2:Ag:Cu on *P. putida* growth, determined by the Koch's method (plating on solid nutrient medium and counting of the survived cells), (a), where the black line presents the bacterial quantity for the control without TiO_2:Ag:Cu film and the red line represents the quantity of survived cells in sample with TiO_2:Ag:Cu film, both in logarithm scale and (b) the results by the optical density measurements at λ=610 nm. The as deposited TiO_2:Ag:Cu films demonstrate no bacterial growth after the 2nd hour from the beginning. This effect on bacterial development continues till the end of the experiment at the 24th hour. In the experiment with the control (the same nutrient medium and bacteria CFU/ml as in the experimental sample but without TiO_2:Ag:Cu thin film) the bacterial growth increases up to the 10th hour from the beginning and after that it decreases till the end of the experiment at the 24th hour. This can be explained with the depletion of the nutrient medium. The same trend is observed in the results obtained by the method of optical density measurements. The both methods which have different sensitivity level demonstrate similar trends in inhibition of the development of bacteria in presence of TiO_2:Ag:Cu the films and prove the bactericidal effect of the as deposited TiO_2:Ag:Cu thin films on *P. putida*.

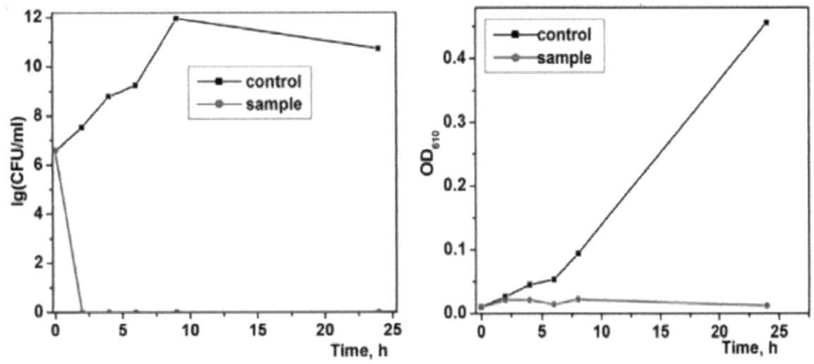

(a) (b)

Figure 2. Antimicrobial effect of as deposited thin films TiO_2:Ag:Cu on bacterial growth of *P. putida,* (a) results by the Koch's method, black line – control and red line – sample, and (b) results by the optical density measurements at λ=610 nm.

Figure 3 demonstrates the antimicrobial effect of annealed thin films TiO_2:Ag:Cu on the growth of bacteria *P. putida* by the method of Koch, (a), and optical density measurements, (b), at λ=610 nm.

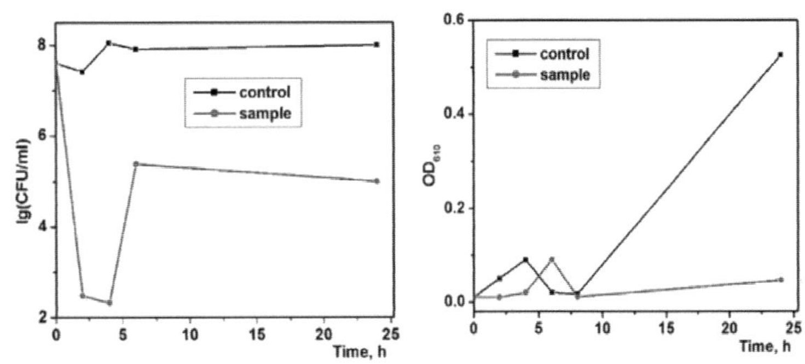

(a) (b)

Figure 3. Antimicrobial effect of annealed thin films TiO_2:Ag:Cu on bacterial growth of *P. putida,* (a) results by the Koch's method, black line - control and red line – sample, and (b) results by the optical density measurements at λ=610 nm.

The both methods of study the antibacterial effect on *P. putida* demonstrate inhibition of the bacterial growth from the beginning till the end of the experiment at the 24th hour. The annealed samples demonstrate weaker inhibition effect on the development of *P. putida* by the both methods in comparison with the as deposited films.

Figure 4 displays the results for study of the antibacterial effect of as deposited thin films TiO_2:Ag:Cu on the growth of *E. coli*. The as deposited thin film TiO_2:Ag:Cu demonstrates only bacteriostatic effect on the development of *E. coli*. The results obtained by the Koch's method (figure 4 (a)) show inhibition up to the 6th hour. The bacteria of the sample with thin film TiO_2:Ag:Cu have an increase in the growth measured as lg CFU/ml after the 6th hour due to nutrient substances evolved from the cells lysis and remained nutrient medium during the inhibition phase.

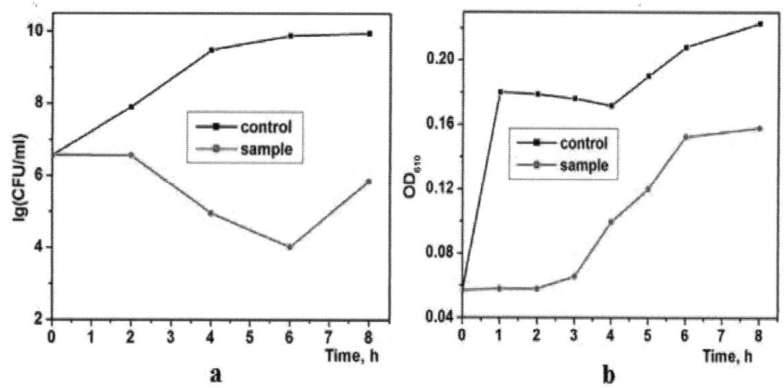

Figure 4. Antimicrobial effect of as deposited thin films TiO_2:Ag:Cu on bacterial growth of *E. coli,* (a) results by the Koch's method, black line – control and red line – sample, and (b) results by the optical density measurements at λ= 610 nm.

The control (without TiO_2:Ag:Cu thin film) demonstrates a decrease in the growth because of the depletion of nutrient medium after the exponential and stationary phase of bacterial growth (up to the 8th hour) obtained by the Koch's method. There is the same trend in inhibition of the bacterial growth in the sample till the 3rd h and an increase in bacterial growth after it according to the results obtained by the optical density measurements (figure 4 (b)). This can be explained with the different sensitivity and timing of the Koch's method and the method of the optical density measurement.

In figure 5 are presented the results for the antimicrobial effect of annealed thin films TiO_2:Ag:Cu on the bacterial growth of *E. coli*.

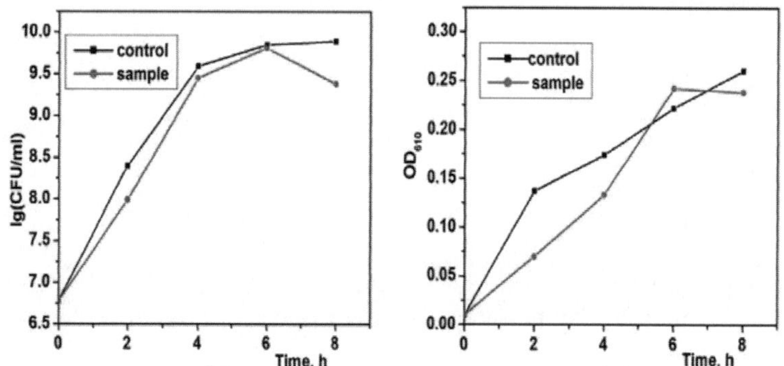

Figure 5. Antimicrobial effect of annealed thin films TiO_2:Ag:Cu on bacterial growth of *E. coli,* (a) results by the Koch's method, black line - control and red line - sample, and (b) results by the optical density measurements at λ=610 nm.

The both methods used for determination of the effect of the annealed films demonstrate no inhibition effect on the bacterial growth and the same trends of increasing the bacterial quantity in comparison with the results of as deposited thin films (figure 5).

The annealed samples demonstrate intensive plasmonic absorption at about 400 nm under illumination but this didn't lead to increase in the antimicrobial effect. This can be explained with the influence of doping Ag and Cu atoms on the process of recombination of the generated electrons under illumination as recombination centres of transition metals inside TiO_2 [13]. The defective states in amorphous TiO_2 deposited by r.f. magnetron sputtering are responsible for photocatalytic inactivity of the film [14]. This leads to loss of electrons from the bulk. Such electron-holes cannot reach the surface of TiO_2:Ag:Cu and cannot react with the attached to the film bacteria for creation of reactive oxygen species (ROS). Thus, small amount of electron-holes generated on the surface of TiO_2:Ag:Cu could participate in formation of ROS. The authors in [15] are established the necessity for the film thickness to be higher than a threshold value for good photocatalytic properties of amorphous TiO_2 doped with transition metals. The mechanism of antibacterial effect of as deposited thin films TiO_2:Ag:Cu maybe is due to metal ions toxicity to bacteria or to mechanistic demolition of cell wall by small nanoparticles, but is not due to formation of reactive oxygen species as established in [16].

4. Conclusion

The antimicrobial effect of TiO_2 doped with Ag and Cu on Gram-negative bacteria *Escherichia coli* and *Pseudomonas putida* is studied. The as deposited thin films TiO_2:Ag:Cu are amorphous and have band gap energy of 3.56 eV little higher than the band gap of crystalline anatase TiO_2 which can be explained with the quantum effect of the granular structure of r.f. magnetron sputtered films. The annealed samples have band gap of 2.52 eV due to formation of donor levels from Ag and Cu atoms near the bottom of the conduction band of the films. The as deposited TiO_2:Ag:Cu thin films demonstrate bactericidal effect on *P. putida* and an inhibition effect (up to the 6th hour) on *E. coli*, which is stronger in comparison with the annealed samples. The both methods of study the antimicrobial effect show the same trends of the bacterial growth independently of their different sensitivity. It is necessary to elucidate the mechanism of antibacterial action of as deposited thin films TiO_2:Ag:Cu. For the annealed thin films TiO_2:Ag:Cu we can conclude that it is necessary to elucidate the possibility for improvement of antibacterial effect and their plasmonic properties through variation of Ag and Cu content as dopants in TiO_2 thin films.

Acknowledgements

This study was conducted with the support of the Department of General and Industrial Microbiology in Faculty of Biology of the Sofia University "St. Kl. Ohridski" and the Central Laboratory of Solar Energy and New Energy Sources, Bulgarian Academy of Sciences. The authors are thankful to INERA project "Research and Innovation Capacity Strengthening of ISSP-BAS in Multifunctional Nanostructures", REGPOT-2012-2013-1 NMP for the financial support for our participation in the conference Velingrad, 2016.

References

[1] Huang J, Liu Y, Lu L and Li L 2012 *Res. Chem. Intermed.* **38** 487
[2] Fujishima A, Zhang X and Tryk D 2008 *Surf. Sci. Reports* **63** 515
[3] Hajipour M, Fromm K, Ashkarran A, de Aberasturi D, de Larramendi I, Rojo T, Serpooshan V, Parak W, and Mahmoudi M 2012 *Trends in Biotechnology* **30** 499
[4] Behbahani M, Babapour M, Amini M, Sadeghi O, Bagheri A, Salarian M and Rafiee B 2013 *American J. Anal. Chem.* **4** 90
[5] Savage N and Diallo M 2005 *Journal of Nanoparticle Res.* **7** 331
[6] Akhavan O 2009 *J. Coll. and Int. Sci.* **336** 117
[7] Foster H A, Sheel D W, Sheel P, Evans P, Varghese, Rutschke N, Yates H M 2010 *J. Photochem.Photobiol. A: Chemistry* **216** 283
[8] Arellano U, Asomoza M and Ramirez F 2011 *J. Photochem.and Photobiol. A:Chemistry* **222** 159
[9] Pankove J 1971 *Optical Processes in Semiconductors* (Prentice-Hall, Inc., New Jersey)
[10] Tauc J 1968 *Mater. Res. Bull.* **3** 37
[11] Linsebigler A L, Lu G, Yates J T Jr. 1995 *Chem. Rev.* **95** 735
[12] Kreibig U and Vollmer M 1995 *Optical Properties of Metal Clusters* (Springer, Berlin, vol. 25)
[13] Mu W, Herrmann J and Pichat P 1989 *Catal. Lett.* **3** 73

[14] Nakamura M, Toru A, Hatanaka Y, Korzec D and Engemann J 2001 *J. Mater. Res.* **16** 621
[15] Huang J, Liu Y, Lu L and Li L 2012 *Res. Chem. Intermed.* **38** 487
[16] Quiñones-Jurado Z V, Waldo-Mendoza M A, Aguilera-Bandin H M, Villabona-Leal E G, Cervantes-González E, Pérez E 2014 *Mater. Sci. Applic.* **5** 895

Antimicrobial effect of Al_2O_3, Ag and Al_2O_3/Ag thin films on Escherichia coli and Pseudomonas putida

O Angelov[1], D Stoyanova[1,2], I Ivanova[2] and S Todorova[3]

[1]Central Laboratory of Solar Energy and New Energy Sources, Bulgarian Academy of Sciences, 72 Tzarigradsko Chaussee, 1784 Sofia, Bulgaria
[2]Department of "General and Industrial Microbiology", Faculty of Biology, Sofia University "St. Kl. Ohridski", 8 D. Tzankov, 1164 Sofia, Bulgaria
[3]Department of Biotechnology and Food Technology, University of Russe "A. Kanchev", 42 "Aprilsko vastanie" blvd., Razgrad

E-mail: oangelov@phys.bas.bg

Abstract. The influence of Al_2O_3, Ag and Al_2O_3/Ag thin films on bacterial growth of Gram-negative bacteria *Pseudomonas putida* and *Escherichia coli* is studied. The nanostructured thin films are deposited on glass substrates without intentional heating through r.f. magnetron sputtering in Ar atmosphere of Al_2O_3 and Ag targets or through sequential sputtering of Al_2O_3 and Ag targets, respectively. The individual Ag thin films (thickness 8 nm) have a weak bacteriostatic effect on *Escherichia coli* expressed as an extended adaptive phase of the bacteria up to 5 hours from the beginning of the experiment, but the final effect is only 10 times lower bacterial density than in the control. The individual Al_2O_3 film (20 nm) has no antibacterial effect against two strains *E. coli* - industrial and pathogenic. The Al_2O_3/Ag bilayer films (Al_2O_3 20 nm/Ag 8 nm) have strong bactericidal effect on *Pseudomonas putida* and demonstrate an effective time of disinfection for 2 hours. The individual films Al_2O_3 and Ag have not pronounced antibacterial effect on *Pseudomonas putida*. A synergistic effect of Al_2O_3/Ag bilayer films in formation of oxidative species on the surface in contact with the bacterial suspension could be a reason for their antimicrobial effect on *E. coli* and *P. putida*.

1. Introduction

The environmental changes or frequent misuse of antibiotics in recent years influence the development of different mutant microbial species resistant to drug treatment. Among the various attempts to surmount the bacterial resistance metal oxide nanoparticles provide encouraging results. It is necessary to study metal oxides, which are chemically stable in different media, exhibiting surface electrostatic charge effect, etc. These properties are promising in bacterial neutralization [1]. The antibacterial activity of nanoparticles depends on their size, surface to volume ratio, environmental pH, etc. The nanoparticles can be prepared in size much smaller than the bacteria. They can penetrate into the cell and attach to important structures inside the bacteria and DNA. Different reactions with bacterial cells are established and are studied for elucidation of nanoparticles - bacteria interaction [2, 3]. The great interest in metal oxide nanoparticles bactericidal effect encounters different challenges concerning to their nature, in particular, their segregation in the solutions, which requires additional care for homogenization of the solutions [4]. Static films of metal oxide are attractive for application in neonatal incubator walls, in sterile rooms, on wound dressing, etc., as an antimicrobial film. Antiseptic or antimicrobial properties of thin films ZnO and Al_2O_3 deposited with Atomic Layer Deposition (ALD) method on commercial polymer films typically used for packaging purposes are studied in [5]. Recently, quantum dots of CuO are deposited on glass surface by low-temperature solution process formulated as paint to prepare thin film on glass [6]. The as-coated thin film, thickness about 120 nm, demonstrated high contact bacterial killing effect against *Staphylococcus aureus* and *Escherichia coli*. The efficient antibacterial activity is due to intracellular reactive oxygen species (ROS) generated by the CuO quantum dots, which interact with the bacterial cells. This leads to an oxidative attack and finally results in bacterial cell death. The authors propose both contact killing and/or copper ion release killing mechanisms for the antibiofilm activity.

This article presents results for study of the antimicrobial effect of thin films Al_2O_3, Ag and Al_2O_3/Ag on bacterial growth of Gram-negative bacteria *Pseudomonas putida* and *Escherichia coli*.

2. Experimental

Thin films Al_2O_3, Ag and Al_2O_3/Ag are deposited on glass substrates without intentional heating during the deposition by r.f. magnetron sputtering of Al_2O_3 and Ag targets or by sequential sputtering of Al_2O_3 and Ag targets, respectively. The sputtering atmosphere in the case of Al_2O_3 is Ar (2 Pa) and in the case of Ag is Ar (0.2 Pa). The sputtering power of Al_2O_3 and Ag thin film is 50 W and 30 W, respectively. The thickness of Al_2O_3 films is 20 nm and this of Ag is 8 nm, measured with profilometer Taylor Hobson.

The microorganisms in this study were supplied by the National Bank for Industrial Microorganisms and Cell Cultures and National Center of Infectious and Parasitic Deseases: *E. coli* 3548 (ATCC 10536) by NBIMCC(industrial strain), *E. coli* ATCC 25922 by NCIPD (pathogen strain) and *Pseudomonas putida* 1090 (ATCC 12633) by NBIMCC.

The method for study of thin film antimicrobial effect is inhibition of bacterial growth in dynamic regime. The toxic effect was determined through the classical Koch's method (plating on solid nutrient medium and counting of survived cells). Two strains of test bacteria *E. coli* were used - industrial and clinical isolates and one strain of *Pseudomonas putida*. The control sample didn't contain any of the studied thin films. The inoculum quantity and nutrient medium in all variants was constant at application on the studied films.

3. Results and discussion

Figure 1 demonstrates the results for the antimicrobial effect of thin films Al_2O_3/Ag on the bacterial growth of *Ps. putida* and the both strains of *E. coli*. The bacteria *Ps. putida* were destroyed thoroughly at the second hour from the beginning of the experiment and did not recover till the 24th hour.

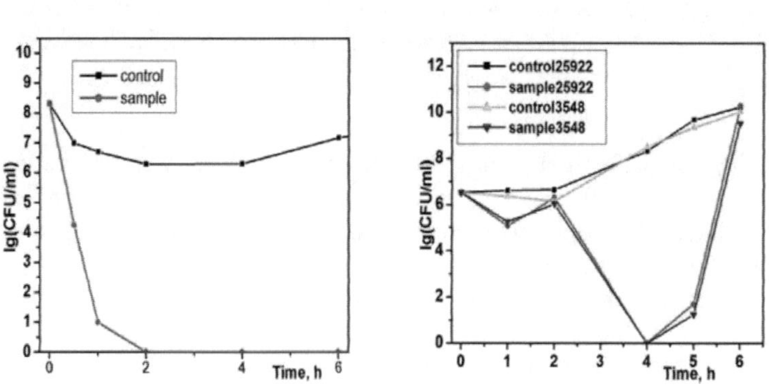

(a) (b)

Figure 1. Antimicrobial effect of thin films of Ag/Al_2O_3 on bacterial growth of (a) *Ps. putida* and (b) *E. coli,* pathogen strain (25922) and industrial strain (3548).

The control variant at the same time demonstrated average growth rate of 1,5 x10^6 cells/hour. From these results an effective disinfection time of 2 hours for *Pseudomonas putida* in presence of Al_2O_3/Ag thin film can be determined. The two figures are represented up to the 6th h because during this period the main processes take place.

The industrial and pathogenic strains of *Escherichia coli* also demonstrate retention in the growth in presence of Al_2O_3/Ag film figure 1 (b). Strong decrease in the bacterial quantity was observed till the fourth hour – the bacterial colonies were not detected on the nutrient media at this hour by the method of Koch, but only an hour later some colonies appeared and started to grow rapidly. Obviously, a bacterial transformation has occurred and a new resistant mutant colonized the experimental medium.

This result can be explained by neutralization of the dissolved nanoparticles from chelating agents emitted from the bacteria and the rich protein content of nutrient medium. The culture medium for *Escherichia coli* is rich in organic (meat peptone broth) but the nutrient medium of *Pseudomonas putida* is entirely synthetic. Obviously, this plays an important role on the antibacterial effect of the thin films. Other fact which could influence the different effects of the thin films on the bacterial growth is that the bacteria belong to different genera, although they have a similar structure of the cell wall.

It can be concluded, that the inhibition effect of the studied thin films on the bacterial growth depends on the composition of the thin films and the test method. The two layer structure Al_2O_3/Ag has an inhibition effect on *E. coli* during the first four hours. Similar structure but prepared by impregnation method is studied in [2]. The authors used scavengers for reactive oxygen species (ROS) on the surface of Al_2O_3/Ag structure and they established no inhibition of the bacterial growth in comparison with the pure Al_2O_3/Ag structure. This indicates formation of ROS when the Al_2O_3/Ag structure is in contact with the bacterial suspension. These ROS can be a reason for the bacteriostatic effect. As can be seen from figure 3, the both strains of *E. coli* have shown some retardation in its development during the first 3- 5 hours, but after that due to their adaptation to the environment they begin to grow exponentially till the 24th hour. The measurements at the 24th hour are not represented in the figures for clarity of the cells reaction in the early hours of the bacterial culture development.

Figure 2 displays the growth of *Ps. putida* and *E. coli* in presence of thin film Al_2O_3. An inhibition effect in bacterial development of *Ps. putida* in the experiment with control and sample (in presence of Al_2O_3 film) is established between the 2th and 4th hour from the beginning and after the 6th hour the bacterial growth increases till the 24th hour. A prolonged lag-phase till the second hour in the development of *E. coli* is observed (figure 2 (b)). No significant differences between the both strains of *E. coli* – clinical and industrial - were established.

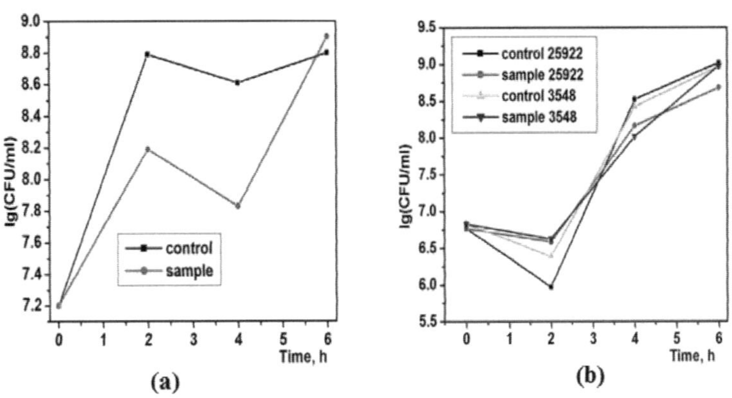

Figure 2. Antimicrobial effect of thin films of Al_2O_3 on bacterial growth of (a) *Ps. putida* and (b) *E. coli,* pathogen strain (25922) and industrial strain (3548).

Figure 3 displays the bacterial development in presence of Ag film.

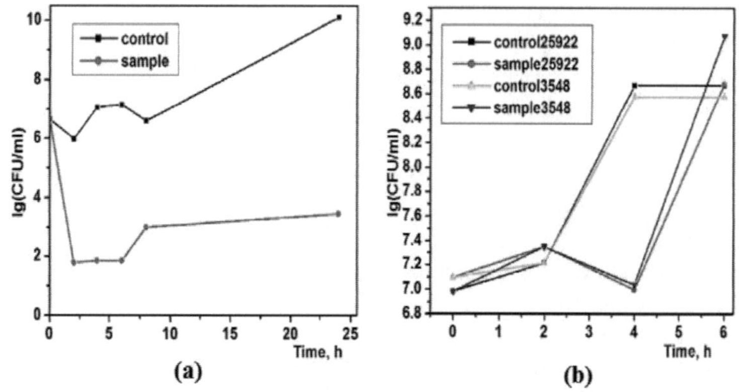

Figure 3. Antimicrobial effect of thin films Ag on the bacterial growth of (a) *Ps. putida* and (b) *E. coli* industrial (3548) and pathogen (25922) strains.

Retention in the growth during the first 4-5 hours was observed. As the nutrient medium for the inoculum activation and the experiment had the same content and the inoculum was in exponential phase, the delay of bacterial growth was probably due to the influence of the Ag thin film only. It is expected, that metal nanoparticles were eluted from the thin film [4] and have interacted with the cell wall, which decreased the bacteria multiplication rate. The authors of [7] established that in Ag film at ambient conditions exist oxidized silver atoms. The oxide layer on the Ag film could initiate release of Ag^+, which has antibacterial effect as Ag atoms [8]. Other reason for the antimicrobial activity at the first hours of the experiment could be the morphology of the film – the existence of nanoparticles with different size. It is established [8] a fast release first of small nanoparticles and after that the film is restructured due to formation of film with big nanoparticles which have a lower rate of dissolution and the inhibition effect of the Ag film on the bacterial growth lowers, so the bacteria continue to develop in dependence on the remained nutrient medium [9]. The Ag thin films in this study are deposited on glass substrate by r.f. magnetron sputtering and have columnar structure with grains different in their size [10].

The Ag thin film proved a weak inhibition effect on *E. coli* up to the 4th hour of the experiment, but at the 6th hour the bacterial density in the control and in the sample were similar. This result probably shows that at low concentrations the silver nanoparticles or their ions [4] have a weak effect on the development of *E. coli* and after the 4th hour they adapt to nutrient medium and show exponential growth of cells in the samples (not represented in the figures). According to the displayed graphs the difference between the both bacterial strains of *E. coli* (industrial and pathogen) is so insignificant that it is within the error.

4. Conclusion

Antimicrobial effect of Al_2O_3, Ag and Al_2O_3/Ag thin films on bacterial growth of Gram-negative bacteria *Pseudomonas putida* and *Escherichia coli* is studied. The antimicrobial effect of thin films was determined as inhibition of bacterial growth in periodic culture. The toxic effect was determined through the classical Koch's method. The Ag thin films have a weak inhibition effect on *Escherichia coli* expressed as an extended adaptive phase of the bacteria up to 5 hours from the beginning of the experiment, but the final effect at the 24th hour is only 10 times lower bacterial density than in the control. This could be due to release of Ag atoms and Ag^+ ions from the films faster at the beginning of the experiment. The Al_2O_3 thin film has no pronounced inhibition effect against the two strains of *E. coli* - industrial and pathogenic. The individual films Al_2O_3 and Ag have not pronounced antibacterial effect on *Pseudomonas putida* too. The Al_2O_3/Ag bilayer films (Al_2O_3 20 nm/Ag 8 nm) have strong bactericidal effect on *Pseudomonas putida* with an effective time of disinfection for 2 hours. A synergistic effect of Al_2O_3/Ag bilayer films in formation of oxidative species on the surface in contact with the bacterial suspension could be a reason for their antimicrobial effect on *E. coli* and *P. putida.*

Acknowledgements
This study was conducted with the support of the Department of General and Industrial Microbiology in Faculty of Biology of the Sofia University "St. Kl. Ohridski" and the Central Laboratory of Solar Energy and New Energy Sources, Bulgarian Academy of Sciences. The authors are thankful to INERA project "Research and Innovation Capacity Strengthening of ISSP-BAS in Multifunctional Nanostructures", REGPOT-2012-2013-1 NMP for the financial support for our participation in the conference Velingrad, 2016.

References

[1] Azam A, Ahmed A, Oves M, Khan M, Habib S and Memmic A 2012 *Inter. J. Nanomedicine* **7** 6003
[2] Chang Q, Yan L, Chen M, He H and Qu J 2007 *Langmuir* **23** 11197
[3] Ansari M, Khan H, Khan A, Cameotra S, Saquib Q and Musarrat J 2013 *J. Appl. Microbiol.* **116** 772
[4] Agnihotri S, Mukherjiabc S and Mukherji S 2013 *Nanoscale* **5** 7328
[5] Vähä-Nissi M, Pitkänen M , Salo E, Kenttä E , Tanskanen A, Sajavaara T, Putkonen M, Sievänen J , Sneck A, Rättö M, Karppinen M and Harlin A 2014 *Thin Solid Films* **562** 331
[6] Tripathy N, Ahmad R, Bang S H, Khang G, Min J and Hahn Y B 2016 *ACS Appl. Mater. Interfaces*, Article ASAP, **DOI:** 10.1021/acsami.6b04494
[7] Schmidt M, Masson A and Brechignac C 2003 *Phys. Rev. Lett.* **91** 243401
[8] Sotiriou G A, Meyer A, Knijnenburg J T N, Panke S and Pratsinis S E 2012 *Langmuir* **28** 15929
[9] Nagy A, Harrison A, Sabbani S, Munson R S, Jr, Dutta P K and Waldman W J, 2011 *Int. J. Nanomed.* **6** 1833
[10] Lovchinov K, Petrov M, O Angelov O, Nichev H, Karashanova D and Dimova-Malinovska D 2014 *J. Phys.: Conf. Ser.* **514** 012012

Journal of Physics: Conference Series **764** (2016) 012016 doi:10.1088/1742-6596/764/1/012016

Electrical characteristics of multilayered HfO_2 – Al_2O_3 charge trapping stacks deposited by ALD

D Spassov[1], A Paskaleva[1], E Guziewicz[2], G Luka[2], T AKrajewski[2], K Kopalko[2], A Wierzbicka[2] and B Blagoev[1]

[1] Institute of Solid State Physics, Bulgarian Academy of Sciences, Tzarigradsko Chaussee 72, Sofia 1734, Bulgaria
[2] Institute of Physics, Polish Academy of Sciences, Al. Lotników 32/46, 02-668 Warsaw, Poland

E-mail: d.spassov@issp.bas.bg

Abstract. Electrical and charge trapping properties of atomic layer deposited HfO_2-Al_2O_3 multilayer stacks with two different Al_2O_3 sublayer thicknesses were investigated regarding their implementation in charge trapping non-volatile memories. The effect of post deposition annealing in oxygen at 600°C is also studied. The decreasing Al_2O_3 thickness increases the stack's dielectric constant and the density of the initial positive oxide charge. The initial oxide charge increases after annealing to $\sim 6\times 10^{12}$ cm^{-2} and changes its sign to negative for the stacks with thicker Al_2O_3. The annealing enhances the dielectric constant of the stacks and reduces their thickness preserving the amorphous status. Nevertheless the annealing is not beneficial for the stacks with thicker Al_2O_3 as it considerably increases leakage currents. Conduction mechanisms in stacks were considered in terms of hopping conduction at low electric fields, and Fowler-Nordheim tunnelling, Schottky emission and Poole-Frenkel effect at higher ones. Maximum memory windows of about 12 and 16V were obtained for the as-grown structures with higher and lower Al_2O_3 content, respectively. In latter case additional improvement (the memory window increase up to 23V) is achieved by the annealing.

1. Introduction

The high-*k* dielectrics based on transition metals oxides have attracted substantial interest for the past 20 years as a solution to scaling issues of advanced ultra-high density microelectronic devices [1,2]. The properties of the high-*k* materials, the fabrication techniques and conditions have been tailored to obtain ultra-thin films allowing the replacement of thermal SiO_2 and SiON as a gate dielectrics in filed effect transistors (FET) and storage capacitors of dynamic random access memories. The intensive investigations revealed that these materials are prone to a high density of bulk traps. A lot of efforts have been paid to overcome this problem. The presence of bulk traps in the high-*k* oxides, although considered as a drawback from advanced FET and storage capacitors point of view, can be regarded as quite interesting and the efficient trapping and retention of charge to be used in other micro(nano)electronic device applications.

Non-volatile memory devices and in particular flash memories became extremely widely used recently. The flash memory cell is built on floating gate FET. The dimensions of the floating gate cell have been shrunk continuously to ensure density growth pace of ~ 2× every 2 years [3]. As the scaling of floating gate memory approaches the innate limitations of this technology several solutions based on

Journal of Physics: Conference Series **764** (2016) 012016 doi:10.1088/1742-6596/764/1/012016

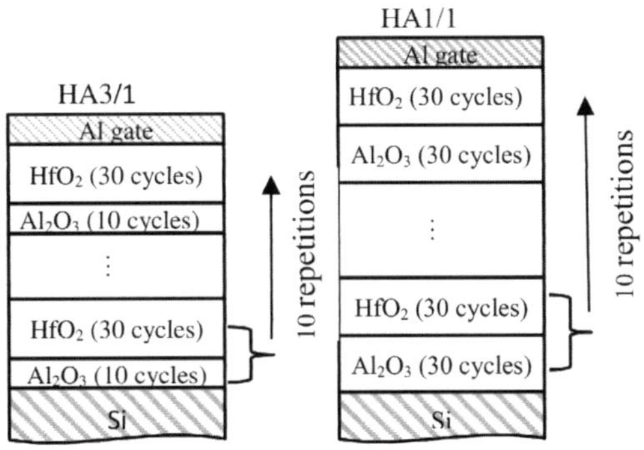

Figure 1. Schematic cross section of the investigated structures.

different memory effects have been proposed [4]. Among them charge-trapping memory (CTM) has received a lot of attention for its full compatibility with the current floating gate technological process. The charge in CTM is stored in spatially discrete traps in the bulk of the dielectric layer instead of the floating gate. The trap rich materials as high-*k* metal oxides are excellent charge trapping (CT) medium, moreover the future flash devices will require high perimittivity gate dialectics [4] as well.

HfO_2 is one of the most heavily investigated high-*k* dielectrics. The maturity of HfO_2 technological process ensured its introduction in Intel Penryn and Samsung A7 processors. Furthermore, some recently published results demonstrate that HfO_2 can be implemented as CT layer in future flash memories [4,5]. It was also proposed that the CT characteristics of HfO_2 could be substantially boosted by introduction of Al into its matrix, by creation of HfO_2/Al_2O_3 laminated stacks [6,7]. The effect of Al into HfO_2, however, is far from well understood and more studies clarifying the role of Al are needed. In this work we report electrical characteristics of multilayer HfO_2/Al_2O_3 stacks prepared by atomic layer deposition and their dependence on the Al_2O_3 content and oxygen annealing. The charge trapping in the structures was also addressed.

2. Experimental procedure

Multilayered HfO_2-Al_2O_3 CT stacks were deposited on p-type (100) Si wafers with resistivity 6–8 Ωcm. Prior to the deposition the substrates were chemically cleaned. Two types of stacks with different Al_2O_3 content (thickness of sublayers) were grown in Savannah-100 ALD system at a temperature of 135°C. Trimethylaluminum precursor was used for Al_2O_3 deposition and tetrakis(dimethylamido)hafnium was used for HfO_2. The oxidant was deionized H_2O. First Al_2O_3 films of 10 or 30 cycles were grown on Si followed by 30 cycles of HfO_2 deposition. The Al_2O_3-HfO_2 deposition sequence was repeated 10 times, resulting in multilayered stacks with identical thickness of HfO_2 and variant Al_2O_3 one, referred to as HA3/1 and HA1/1 in case of 10 and 30 Al_2O_3 cycles respectively (figure 1). Part of the samples received Post Deposition Annealing (PDA) in dry O_2 at 600 °C for 20 min. The overall thickness, *d* was evaluated by Woolman M2000D ellipsometer in spectral range of 500–1000 nm. *d* of as-grown HA3/1 stacks is 49.2 nm and 65 nm for the HA1/1 ones. PDA reduces the thickness of the stacks. The reduction of *d* is stronger (7.7 %) for the thinner HA3/1 samples while for HA1/1 stacks the decrease is ~5.4%, resulting in *d*=45.4 nm (HA3/1) and 61.5nm (HA1/1). The crystallinity of the structures and their surface morphology were examined by X-Ray Diffraction (XRD) and Atomic Force Microscopy (AFM), respectively. MOS capacitors with Al top (gate) and backside electrodes fabricated by sputtering were used for the electrical characterization. The square Al top electrodes (area of 10^{-4} cm^2) were patterned by photolithography. All electrical measurements were carried out in dark chamber employing Agilent 4980A LCR meter for capacitance-voltage (*C-V*) characteristics, and Keithley 236 SMU for the current-voltage (*I-V*) ones.

3. Results and discussion

Both types of as-grown stacks are amorphous as revealed by XRD analysis. The spectra of HA3/1 and HA1/1 layers are identical showing only features due to crystalline Si substrate. (XRD data is illustrated for HA3/1 structure in figure 2(a)). The PDA does not change the amorphous status of the films, despite the ellipsometrycally obtained thickness reduction, i.e. PDA densifies stacks but their amorphous structure is preserved. The surface morphology of HA3/1 stacks is presented in figure 2(b) and c before and after PDA, respectively. Noticeable differences between the two types of as-grown CT structures were not observed (the average roughness, R_a of HA3/1 layers is 0.433 nm and it is slightly lower 0.407nm for HA1/1). PDA, however, slightly smoothens the surface (R_a = 0.426 and 0.334 nm for HA3/1 and HA1/1, respectively).
The initial *C-V* characteristics of the as-grown and annealed structures measured in the frequency range f = 1kHz to 1MHz are presented in figure 3. The curves were obtained in a narrow applied voltage range

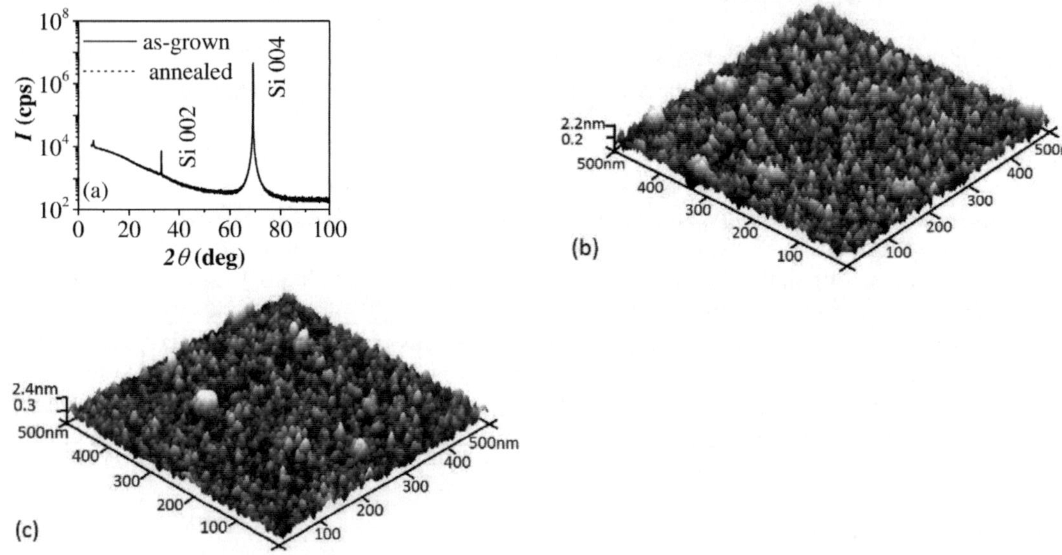

Figure 2. XRD spectra (a) and AFM data for HA3/1 samples before (b) and after annealing (c).

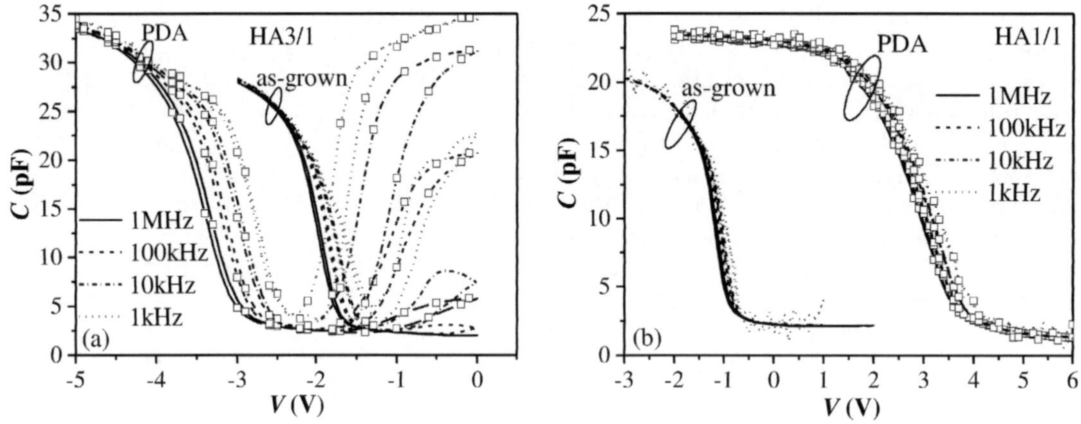

Figure 3. Initial *C-V* curves obtained at different frequencies of HA3/1 (a) and HA1/1 (b) stacks before and after annealing, (designated by □ symbol).

in which the charge trapping is negligible. The hysteresis of the 1MHz *C-V* curves is ~0.05V. As evidenced from figure 3 the capacitance in accumulation does not depend on f. The stack capacitance

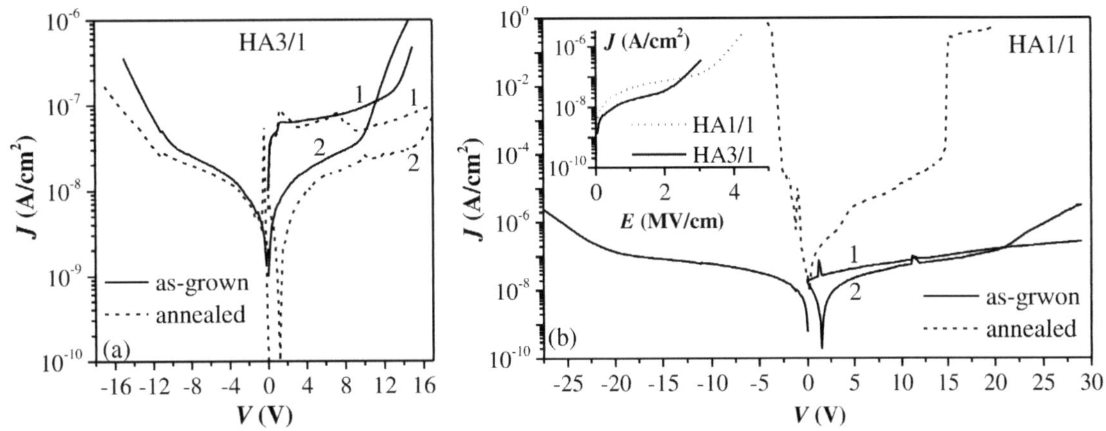

Figure 4. Leakage currents of stacked HA3/1 (a) and HA1/1 (b) capacitors before and after PDA. 1 and 2 denote the first and the second consecutive measurements in $+V$. Inset: J in negative biases of the as-grown stacks vs. applied electric field.

was evaluated by Kar's technique [8] and was further used to determine the effective dielectric constant ε_{eff}. ε_{eff} of the as-grown HA3/1 and HA1/1 are 17.7 and 15.5, respectively. PDA increases the dielectricconstant of the stacks. ε_{eff} ~19 is obtained for the annealed HA3/1 samples and ~18 for the annealed HA1/1 ones. The increase of ε_{eff} after PDA is in a good agreement with the observed reduction of d and is most likely result of stacks' densification. A significant difference in the initial oxide charge of the stacks Q_{ox} (i.e. the charge present in the structures, prior to any carrier injection) as well as its response to the PDA is found upon variation of Al_2O_3 content. Q_{ox} of both types of as-grown CT layers is positive. The structures with low amount of Al_2O_3, however, demonstrate much higher Q_{ox} up to ~2.7×10^{12} cm^{-2} than capacitors with thicker Al_2O_3 sublayers (~6×10^{11} cm^{-2}). PDA increases the density of Q_{ox}, up to ~6.4×10^{12} for both stacks. But while for HA3/1 samples Q_{ox} preserves its sign and C-V curves after annealing are shifted towards more negative V, in case of HA1/1 the oxide charge is negative, C-V curves are shifted to the positive V (figure 3). The high positive Q_{ox} is responsible for the observed dependence of the inversion capacitance of HA3/1 structures on f. The high positive Q_{ox} inverts the Si surface, so the area outside the capacitor serves as a supply for minority carriers (electrons) into the inversion layer ensuring that their density follows at some extent the measurement signal frequency. As a result the inversion capacitance is higher than C determined by the depletion layer. The effect is better pronounced in the annealed HA3/1 capacitors which have higher Q_{ox}, i.e the inversion is stronger, and C in inversion approaches at 1 kHz the values of C in accumulation (figure 3(a)). The shift of C-V curves and the hysteresis for HA3/1 stacks at $f <$ 1MHz can be attributed to the combined effect of the slow and fast interface states. PDA enhances these effects implying that the treatment generates additional defects near and at the very interface. Although hysteresis of C-V curves is not observed for HA1/1, the decreased slope of the curves after annealing is a clear indication of increased density of fast interface states. The difference in Q_{ox} between HA3/1 and HA1/1 layers is probably related to the negative fixed oxide charge in Al_2O_3 [10] and additional positive and negative volume charges in Al_2O_3 and HfO_2. All these charges depend strongly on d if it is in the nanometer range as well as deposition conditions and thermal treatments and can lead to Q_{ox} variations in a broad range.

Figure 4 shows the leakage current density, J vs. applied gate voltage of the as-grown and annealed structures. Two consecutive curves were measured at $+V$ (indicated with 1 and 2). J-V characteristics for both stack types under both voltage polarities are similar: an initial slow increase of J with V is followed by an abrupt one at high V (an indication of conduction mechanism change). After the negative bias sweep, J at the first measurement under $+V$ is higher (especially in the low voltage range) than J

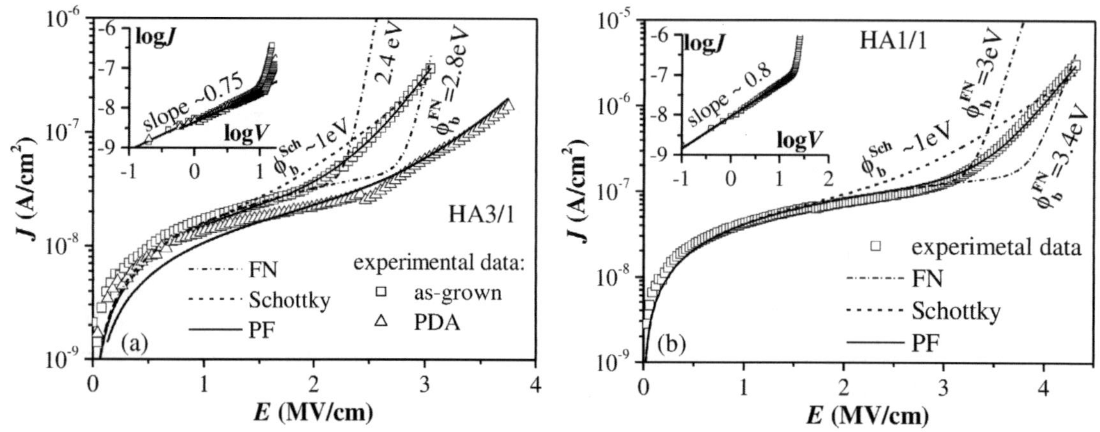

Figure 5. Experimental and simulated *J-E* curves using a combination of Ohmic conduction and Folwler-Nordheim tunnelling, Poole-Frenkel effect and Schottky emission. For FN tunnelling curves corresponding to two barrier heights are presented. (a) HA3/1; (b) HA1/1. Insets: log*J* vs. log*V*.

obtained at the second one. This behaviour is symptomatic for charge trapping phenomena. Under high negative *V* positive charge build-up takes place (via hole injection form Si and/or electron detrapping in the insulator). After the bias polarity switch this positive charge enhances the electric field near the substrate, increasing *J*. Further, the positive charge is compensated and even surmounted by the trapped injected electrons resulting in lower *J* during the second measurement. As seen, the trapping is better pronounced for the stacks with lower Al_2O_3 content, for which the voltage shift at 4×10^{-8} A/cm^2 is ~8 V, while for HA1/1 the maximum voltage shift is ~6 V. The additional *J-V* measurements (not shown here) at +*V* does not result in significant extra voltage shift of the curves, suggesting that at the first *V* sweep the positive charge is released and all electron traps are almost entirely filled. A comparison of the leakage currents of as-grown HA3/1 and HA1/1 capacitors as a function of the applied electric field *E* (defined as *V*/*d*) is shown in the inset of figure 4(b). (*J* only for negative *V* is presented, as for these voltages the Si is in accumulation and the whole *E* is applied across the dielectric.) Both structures exhibit almost identical *J* at low fields, the steeper *J* increase, however, occurs at higher *E* in case of HA1/1 layers. So, thicker Al_2O_3 sublayers provide an improvement of the leakage currents in the high *E* region because larger part of the stacks consists of high band gap dielectric (Al_2O_3). PDA seems to improve to some extent *J* of HA3/1 structures mainly in the high *V* region (the reduction is more distinct at positive biases). The charge trapping feature of *J-V* curves at +*V* is preserved and even enhanced after PDA. The annealing affects *J* of HA1/1 layers in opposite manner: a substantial increase of *J* is detected for both *V* polarities. The lower current of annealed HA1/1 samples at +*V* compared with *J* at -*V* is probably due to the high negative Q_{ox}, preventing the injection of electrons from Si. Because of the large *J* of the annealed HA1/1 layers they will not be further considered.

We briefly discuss the conduction mechanisms of the stacks based on the *J-E* characteristics obtained at room temperature and negative *V*. The following analysis, however, should be regarded only as tentative since generally the temperature dependence of *J* is required for reliable identification of operating conduction mechanisms. Moreover, the charge trapping in the layers can severely affect *E* and complicate the analysis. The insets of figure 5 exhibit log-log plots of *J-V* curves before and after annealing, indicating nearly Ohmic behaviour (resulting from hopping of thermally excited electrons from one trap to another [9]) up to *V* at which *J* abruptly increases. The deviation of the log*J*-log*V* slope from 1.0 (sub-Ohmic behaviour) can be attributed to the charge trapping. Three commonly found conduction mechanisms in high-*k* films: Fowler-Nordheim tunnelling (FN), Schottky emission (Sch) and Poole Frenkel conduction (PF) [9] were invoked to explain the current in the stacks. As evidenced, the best fit of experimental *J-E* curves is obtained by a combination Ohmic conduction and PF

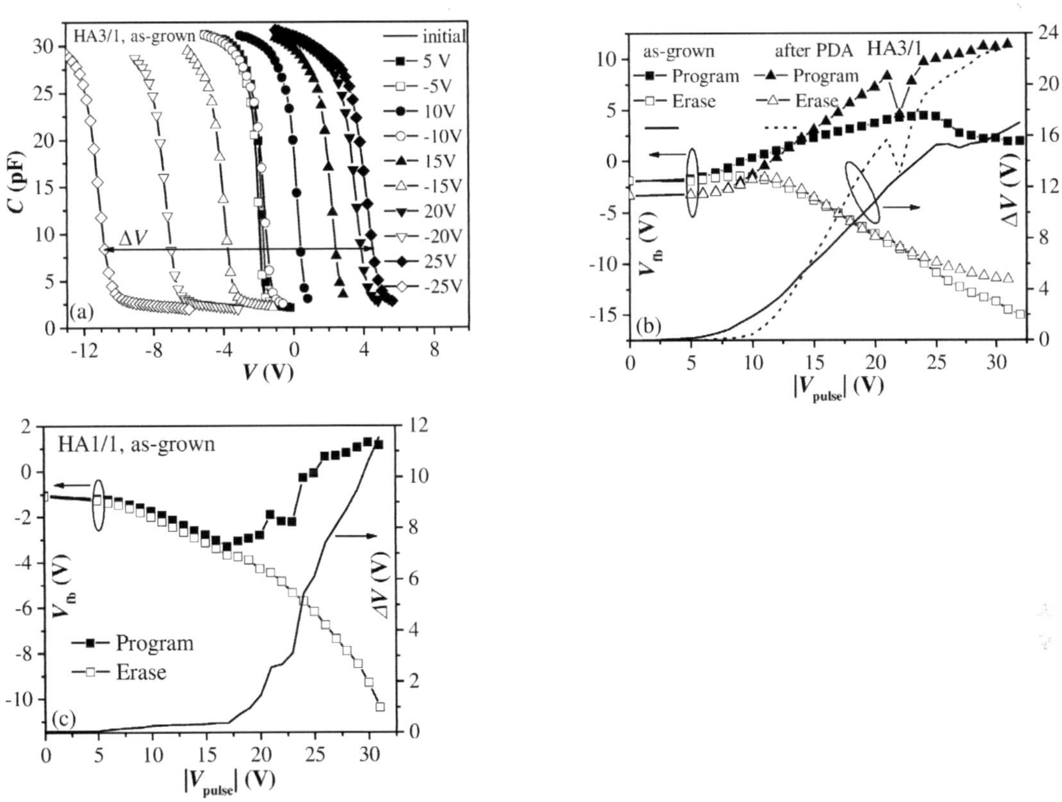

Figure 6. *C-V* curves of as-grown HA3/1 stacks after applying voltage pulses for 1s (a) and evolution of the flatband voltage with the amplitude of the voltage pulse for HA3/1 (b) and HA1/1 samples(c).

mechanism. The values of dynamic dielectric constant K_r (square of refractive index, *n*) and compensation factor *r* used for PF fit are: 3.7 and 1.5, and 3.9 and 2 for as-grown and annealed HA3/1 layers; and 3.4 and 1.2 for the HA1/1 stacks. These values agree very well with the elipsometrically measured *n*, ($n \approx$1.9-1.96 and 1.94-2 for as-grown and annealed HA3/1 stacks, respectively; and 1.82-1.89 for as-grown HA1/1 ones). The combination of Ohmic mechanism and FN fails to describe the high field part of *J*; besides the obtained barrier height, ϕ_b^{FN} is higher than expected for Al/HfO_2 one. A decent interpretation can be made by Schottky emission with barrier height ϕ_b^{Sch}~1 eV and the same K_r as in PF; the fit can be further improved but in all cases leads to values of K_r not consistent with *n*. Taking into account also the symmetry of *J-V* curves in respect to the zero bias we are inclined to attribute the conduction to bulk limited mechanisms – hopping conduction and PF effect.

The charge trapping in the stacks was evaluated by applying square voltage pulses with an amplitude, V_{pulse} from ±5 to ±32V and duration of 1 s, and subsequent measurement of 1 MHz *C-V* curves. The supplying of $+V_{pulse}$ is referred as Program operation while the supplying of $-V_{pulse}$ is noted as Erase one. The shift of the *C-V* characteristics indicative of the trapping in layers is illustrated in figure 6a for as-grown HA3/1 stacks. After positive pulse *C-V* curve shifts to more positive *V* revealing negative charge trapping; the reverse pulse moves the curve backwards. The difference between the flatband voltage, V_{fb} of the curves obtained after Program and Erase operation defines so called memory window, ΔV. (V_{fb} is the applied voltage at which the electric field at Si surface is zero, i.e. there is not Si band bending. V_{fb} is being determined from the *C-V* curves as the voltage at which *C* equals the calculated value of the flatband capacitance [11].) Figure 6(b), (c) summarize the evolution of V_{fb} after Program/Erase operations as well as the value of obtained memory windows as a function of V_{pulse}. Substantial ΔV is obtained for the HA3/1 stacks and PDA additionally extends ΔV form ~17V to ~23V at $|V_{pulse}|$=32V.

The comparison of figure 6(b) and (c) indicates that Al_2O_3 content in stacks affects the evolution of V_{fb} due to the changes in the charge trapping processes and/or different nature of participating traps. For HA1/1 stacks up to $|V_{pulse}|$ ~15 V, ΔV is rather small, and the V_{fb}–V_{pulse} dependence suggest accumulation/generation of positive charge. The capture of electrons occurs at higher $|V_{pulse}|$. The behaviour of HA3/1 capacitors is opposite, (under small V_{pulse} negative charge trapping takes place). The processes of charge trapping/detrapping, retention and endurance characteristics as well as comprehensive analysis of the conduction mechanism will be presented in more details in future work.

4. Conclusion

The dielectric properties of multilayered Al_2O_3-HfO_2 stacks can be effectively modified by the thickness of Al_2O_3 sublayers. The alumina content affects mainly the effective dielectric constant, the initial oxide charge and its response to the thermal treatments, while the impact on the leakage currents is weak. The conduction mechanisms of the stacks is not affected by the Al_2O_3 content and the current is governed by hopping conduction and Poole-Frenkel mechanism. The post deposition annealing densifies the stacks and increases the density of the existing defect states in case of stacks with thinner Al_2O_3 and introduces new type of defects in structures with higher Al_2O_3 content manifested via the change of the oxide charge sign. The higher trap density of stacks with smaller Al_2O_3 content ensures wider memory window, which is further widened by oxygen annealing. Different type of defects participate in the charge trapping and in the conduction in these stacks, since the leakage is improved after annealing. PDA is not suitable for stacks with higher Al_2O_3 content. The results suggest that alumina content should be carefully optimized to obtain maximum performance.

Acknowledgements
The work was carried out in the frame of project INERA REGPOT 316309.

References

[1] Kittl J *et al.* 2009 *Microelectron. Eng.* **7-9** 1789–1795
[2] Atanassova E and Paskaleva A 2007 *Microelectron. Relaib.* **47** 913-923
[3] Zitlaw C 2011 *EE Times* 5/2 http://www.eetimes.com/document.asp?doc_id=1278751
[4] Zhao C, Zhao C, Taylor S and Chalker P 2014 *Materials* **7** 5117
[5] Zhang Y, Shao Y, Lu, X, Zeng M, Zhang Z, Gao X, Zhang X, Liu J and Dai Y 2014 *Appl. Phys. Lett.* **105** 172902
[6] Zhu ,C Huo Z; Xu Z, Zhang M, Wang Q, Liu J, Long S and Liu M 2010 *Appl. Phys. Lett.* **97** 253503
[7] Paskaleva A, Rommel M, Hutzler A, Spassov D and Bauer A 2015 *ACS Appl. Mater. Interfaces* **7** 17032
[8] Kar S 2006 *Thin Solid Films* **504** 178-182
[9] Spassov D, Skeparovski A, Paskaleva and Novkovski N 2016 *Thin Solid Films* in press doi:10.1016/j.tsf.2016.02.064
[10] Simon D, Jordan P, Mikolajick T, and Dirnstorfer I 2015 *ACS Appl. Mater. Interfaces* **7** 28215
[11] Sze S and Ng K Physics of Semiconductor Devices 3rd edition 2007 John Wiley & Sons Hoboken New Jersey p. 205

Energy spectra of a spin-$\frac{1}{2}$ XY spin molecule interacting with a single mode field cavity: Numerical study

H Tonchev, A A Donkov and H Chamati

Institute of Solid State Physics, Bulgarian Academy of Sciences,
72 Tzarigradsko Chaussée, 1784 Sofia, Bulgaria

E-mail: htonchev@issp.bas.bg, aadonkov@issp.bas.bg, chamati@issp.bas.bg

Abstract. In a previous paper [J. Phys.: Conf. Ser. 682 (2016) 012032] we studied analytically the energy spectra of a finite-size spin $\frac{1}{2}$ XY chain (molecule) coupled at an arbitrary spin site to a single mode of an electromagnetic field via the Jaynes-Cummings model. We considered spin rings and open spin molecules with up to 4 spins and an interaction restricted to nearest-neighbours. Here we extend our investigation, addressing numerically the energy spectra of molecules of up to 10 spins with nearest-neighbour or long-range interaction. Furthermore we analyze the behaviour of an invariant operator, constructed by combining the magnetization of the spin-chain and the total number of photons in the system. We found a strong dependence on the number (even or odd) of sites in the molecules. This study is aimed at finding the appropriate combination of the physical parameters that could make the system suitable for use in quantum computations.

1. Introduction

The Jaynes-Cummings [1, 2] model for the interaction of an atom with a quantized electromagnetic field has been used to investigate atom- and ion-field interactions in an ion trap and quantum dots, to study one photon lasers, and for single-photon photodetectors, see for example Refs. [3–6], respectively. Here this model is used in a numerical calculation to obtain the energy spectrum of a magnetic spin chain, or more in tunes with this conference, a magnetic nano-material in the form of a spin molecule, interacting with a laser field.

We consider the easiest, from the standpoint of numerical computation, case of $\frac{1}{2}$ spins of the individual sites in a molecule. The spin-spin interaction has been modelled by the XY model [7], and cases of nearest neighbour and long range interactions are considered. In a previous study [8] we had shown some analytical formulae for the energy spectrum for a 4-site spin molecule, with the numerics we were able to go up to 10-site molecules, but it becomes hard to visualize the full spectra. Due to the specifics of the coupling between the Jaynes-Cummings model and the product states, that diagonalize the spin-spin Hamiltonian, the energy spectrum shows qualitative differences whether we have an odd or an even number of spins in the spin molecule.

In Section 2 we remind the general form of the interaction Hamiltonian used and show some of the spectra for different chain geometries. Then we analyze the obtained spectra in Section 3. We look into the analogy between the invariant that exists in this problem, which helps in classifying the energy levels, with the total magnetization in Section 3.1, then consider the influence of the longer range interaction in Section 3.2, and that of the addition of a site into a chain in Section 3.3, and explore the spectrum in terms of the invariant in Section 3.4, with concluding remarks in Section 4.

2. Laser field coupled to the spin chain

The interaction part, $\widehat{H}$, of the Hamiltonian of our system, which is one mode (one frequency) of a laser field focused on one of the spins of the spin $\frac{1}{2}$ chain of N sites, reads:

$$\widehat{H} = G\left(aS_k^+ + a^\dagger S_k^-\right) - \sum_{i \neq j} 2J_{i,j}\left(S_i^x S_j^x + S_i^y S_j^y\right). \tag{1}$$

The G part is the Jaynes-Cummings (JC) model of the photon $(a, a^\dagger)$ interacting with a two level system (spin S_k on site k), and the second term (factor of 2 is there for convenience) is the XY model of the spin (S_i)-spin (S_j) interaction. We have in mind either nearest-neighbour (NN) case with $J_{ij} = J\delta_{r_j, r_i \pm d}$, or the long-range interactions in the form $J_{ij} = J|r_i - r_j|^{-2}$, with J setting the scale for the spin interaction, r_i, r_j being the coordinates of the spins on the lattice of nearest neighbour distance d, and the indexes i and j enumerate the sites of the chain. For this report we have modelled an open chain geometry, where the ends do not influence each other, and a cyclic arrangement, a spin ring, both with either a NN or a long range spin-spin interaction. For a more detailed description of the notations and the meaning of the terms, as well as the choice of the basis vectors to form the matrix representation of this Hamiltonian, please refer to [8].

Solving the eigenvalue equation det($\widehat{H} - \lambda$ Identity) $= 0$ for a $2^{N+1} \times 2^{N+1}$ matrix gives the energy spectra. These are shown in case of $N = 5$ (top row) and 6 (bottom row) for an open chain (first two columns) and for a spin ring (last three columns) in figure 1. We plot the values of $E(\varphi) = \lambda/\sqrt{G^2 + J^2}$, for a cutout from the parameter space of G and J, such that $G^2 + J^2 = 1,\ \ G > 0$, with the parameter φ defined as $J/G = \tan(\varphi)$. The pure magnetic case of $G = 0$ corresponds to $\varphi = \pi/2$, while the pure JC model falls on $\varphi = 0$ as $J = 0$. The plots for $G < 0$ are simply obtained from the figures by realizing that there is an overall change of sign in the energy, as seen from the equation (1), when $G \to -G, J \to -J$ then $E \to -E$. The total number of energy values for each φ scales as 2^{N+1}, consequently, even with the degeneracy taken into account, the figures for $N = 9$ and 10 become too messy, but show a similar behaviour. Finally, the resulting energy spectra represent the mixture of the photon and the spins, while for an easier comparison with the known results, for example for the a two-level atom with a Jaynes-Cummings model, we will need to project away the photon.

3. Analysis of the energy spectra

3.1. Magnetization and the Invariant for a one photon problem

As it will be seen from the analysis of the plots it is helpful to have an understanding of the magnetic part of the problem, at $\varphi = \frac{\pi}{2}$. First of all, from the two physical entities, the magnetic moment $M_z = \frac{1}{2}\sum_{i=1}^{N} \sigma_i^z$, and the photon fill number, $n_{ph} = a^\dagger a$, we can create an invariant, a combination that commutes with the Hamiltonian of the JC+XY model [9]: $Inv = n_{ph} + M_z,\ [Inv, \widehat{H}] = 0$.

In [8] the states for the electromagnetic field were ordered by first choosing the fill number zero $|0\rangle$, then the fill number one $|1\rangle$ state. If instead we choose first $|1\rangle$, and then $|0\rangle$, while keeping the same order for the spin basis on each site: $|\uparrow\rangle, |\downarrow\rangle$, then this "one" photon problem is represented in terms of spin variables. In this basis the matrix form, in terms of the Pauli matrices, of the photon fill number is similar to introducing the isotopic spin in describing the two states of a nucleon, a neutron and the proton, see e.g. §116 in [10]:

$$n_{ph} \equiv \frac{1}{2} + \frac{1}{2}\sigma_z, \tag{2}$$

the JC interaction [G part in equation (1)] is effectively a XY type interaction, between site K and the extra site 0 chosen to correspond to the photon:

$$aS_k^+ + a^\dagger S_k^- \equiv \sigma^-_{\text{site }0} S_k^+ + \sigma^+_{\text{site }0} S_k^-, \tag{3}$$

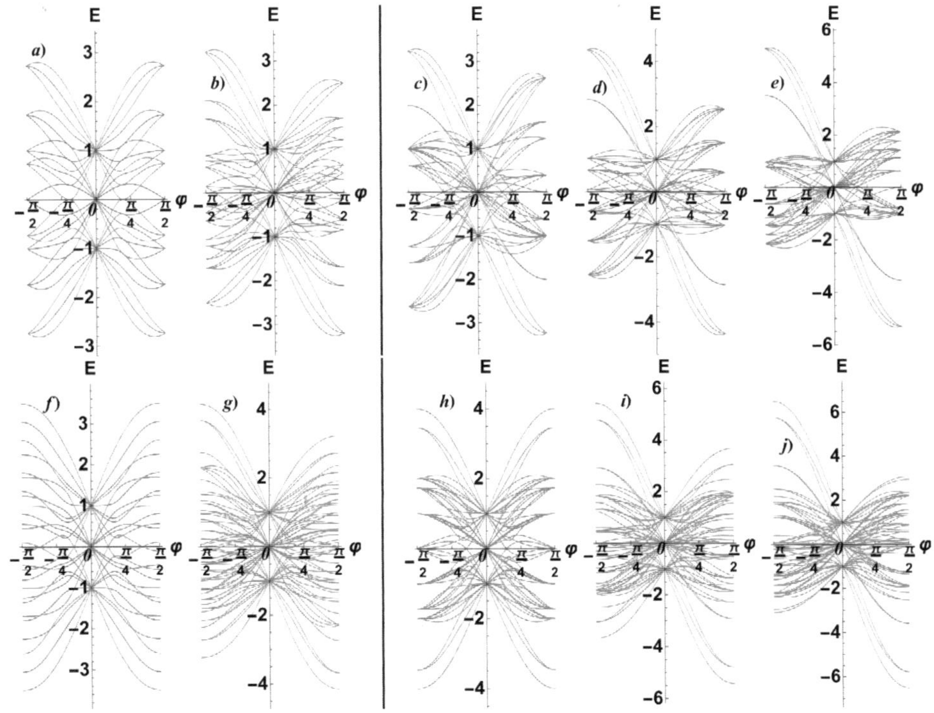

Figure 1. (color online) The energy spectra, in units of $\sqrt{G^2+J^2}$, of the one photon JC-XY model for a 5-spin (top) and a 6-spin (bottom) open chain (two leftmost columns) with a NN (a,f) and long-range (b,g) spin-spin interaction, and for a spin ring arrangement (three rightmost columns), (c,h) NN spin-spin interaction; (d, i) long-range interaction taken along the ring and (e,j) long-range interaction considered to be along the hordes of the spin ring.

and, finally, the invariant is the sum over the chain sites of a spin-1/2 chain with one extra site $i = 0$:

$$Inv = \underbrace{\frac{1}{2} + \frac{\sigma_z}{2}}_{\text{site } 0} + M_z = \frac{1}{2} + \mathscr{M}_z, \text{ where } \mathscr{M}_z \stackrel{def}{=} \frac{1}{2}\sum_{i=0}^{N} \sigma_i^z. \tag{4}$$

We thus have the eigenvalues of Inv : they are immediately obtained by analogy with the familiar problem for the z-component of the angular momentum as $-(N+1)/2, -(N+1)/2+1, \ldots, (N+1)/2$, only shifted by $\frac{1}{2}$.

We have a possible mapping of the problem of N spins and a photon onto a pure spin chain of $N+1$ sites and an anisotropic interaction. Numerically, we can benefit from the advanced packages for solving numerical spin chain problems.

Another useful observation is that for the XY interaction the states with maximum absolute energy are those that have the most number of opposite spin pairs. For example, the $|\uparrow\downarrow\rangle$ and $|\downarrow\uparrow\rangle$ pairs with $M_z = 0$ contribute $\pm J$, and for $J > 0$ the antisymmetric combination of such AFM states combines to give the highest energy level. For 3−spins similarly composed product states $|\uparrow\downarrow\uparrow\rangle$ or $|\downarrow\uparrow\downarrow\rangle$ states have $M_z = \pm\frac{1}{2}$. In case $J < 0$ the symmetric combinations combine for the highest energy level. The fully polarized spin pairs, the FM states $|\uparrow\uparrow\rangle$ with maximum values of M_z, contribute zero, $0J$, to the energy.

For the one photon problem, the presence of the photon states, even at $G = 0$, gives by analogy, that the states with maximal $\mathscr{M}_z$, $|1_{ph}\uparrow\uparrow\rangle\ldots$ and $|0_{ph}\downarrow\downarrow\rangle\ldots$, are the states with an identically equal to zero energy for any G and J values, while the states with $\mathscr{M}_z = 0$ (odd number N) or $\mathscr{M}_z = \pm\frac{1}{2}$ (even number

Journal of Physics: Conference Series **764** (2016) 012017 doi:10.1088/1742-6596/764/1/012017

N) have the maximum energy, which is to be compared with $M_z = 0$ (even number N) or $M_z = \pm\frac{1}{2}$ (odd number N) from above.

3.2. NN vs long range interaction

Open chain: The comparison between the first two columns in figure 1 illustrates the influence of the range of the spin-spin interaction, NN in the first column or long range in the second. The spectrum in figure 1a) is symmetric for reflections about the x and the y axes, and this means, in terms of the interaction, that the NN interaction gives energies that are independent of the sign of either of the parameters G or J, $E^{(a)}(G,J) = E^{(a)}(-G,J) = E^{(a)}(G,-J)$. The spectrum in figure 1b) continues to be independent of the overall sign of the JC part G, as the first and third quadrant energy bands, as well as the 2nd and the 4th, are symmetric to each other, but dependents on the sign of the spin-spin interaction J, as the first and the second quadrant plots are different, $E^{(b)}(G,J) = E^{(b)}(-G,J) \neq E^{(b)}(G,-J)$.

Spin ring: Looking at the last three columns figure 1c), d), and e) we can no longer claim that the spectrum in NN case is independent of the signs of G and J, as this depends on whether the spin ring has an odd (top row),where this is no longer the case, or an even (bottom row) number of spins, for which this is similar to the open chain. This follows from the fact that the XY eigenvalues for the spin ring no longer cluster symmetrically around the zero, $E^{ring}(\pi/2) \neq -E^{ring}(\pi/2)$, even though their weighted average, with the degeneracies taken into account, still is zero. This translates into the 2nd quadrant, by making the spectrum loose its symmetry $E(G,J) \neq E(G,-J)$.

3.3. Even vs Odd number of spin sites

Now we compare the top and bottom row of the figures 1 in order to analyze the effect of introducing an extra spin site into the system. The most striking feature is the more effective lifting of the degeneracy in the odd site spin chain when we move from the purely magnetic case at $\varphi = \pi/2$. Another feature is that the maximum energy values are reached for even N at this magnetic boundary, while for the odd N the maximum is always achieved when a photon is mixed as well, reaching the magnetic boundary only at $N_{odd} \to \infty$.

The more effective lifting of the degeneracy in the case of odd N can be understood when we consider the effect of the JC interaction applied to eigenstates of the problem at $\phi = \pi/2$. In a poor man's perturbation approach, we look only at the maximum energy state. As mentioned at the end of Section 3.1 the corresponding eigenstates are a mixture of the type $|\downarrow\uparrow\downarrow\rangle\ldots$. Then, for example, if the photon interacts with the first spin, the JC part of the Hamiltonian has amplitudes of transition between this and a state written as $|\uparrow\uparrow\downarrow\rangle\ldots$, which means that it has to be diagonalized using more states, than at the point $\phi = \pi/2$, and this bigger matrix has more eigenvalues, thus lifting the degeneracy. For the even numbered, for example $N = 4$, we have the maximum energy state at $\phi = \pi/2$ as a combination of $|\downarrow\uparrow\downarrow\uparrow\rangle$, and the JC part would have matrix elements with smaller range of states and thus there is no lifting of the degeneracy.

3.4. Some spectroscopic analysis using the invariant.

The values of the invariant, in terms of Inv or $\mathcal{M}_z$, can be used to select bands with particular property. For example, let's consider the test case of an open spin chain with 5 or 6 spins with the NN interaction. For $N = 5$, $\mathcal{M}_z = -3,-2,...,2,3$. Similarly, for $N = 6$ we get $\mathcal{M}_z = -7/2,\ldots,7/2$. Some of the graphs for this case are given in figure 2. For $|\mathcal{M}_z^{(N=5)}| = 3$, and $|\mathcal{M}_z^{(N=6)}| = 7/2$, the energy levels are identically equal to zero, and we show the typical eigenstate. In table 1 we give some of the numeric details for this open chain case.

Table 1. Summary of the number of distinct eigenvalues at $\varphi = \pi/2$, the maximum value of the energy and the location of this maximum for each of the nine lengths N of an open spin chain molecule with NN interaction.

N	# of Distinct	E_{max}	φ_{max}
2	3	1	flat
3	3	1.6180	−1.0196
4	9	2.2361	−1.5708
5	9	2.8064	−1.3188
6	27	3.4940	−1.5708
7	27	4.0649	−1.4212
8	58	4.7588	−1.5708
9	91	5.3362	−1.4684
10	256	6.0267	−1.5708

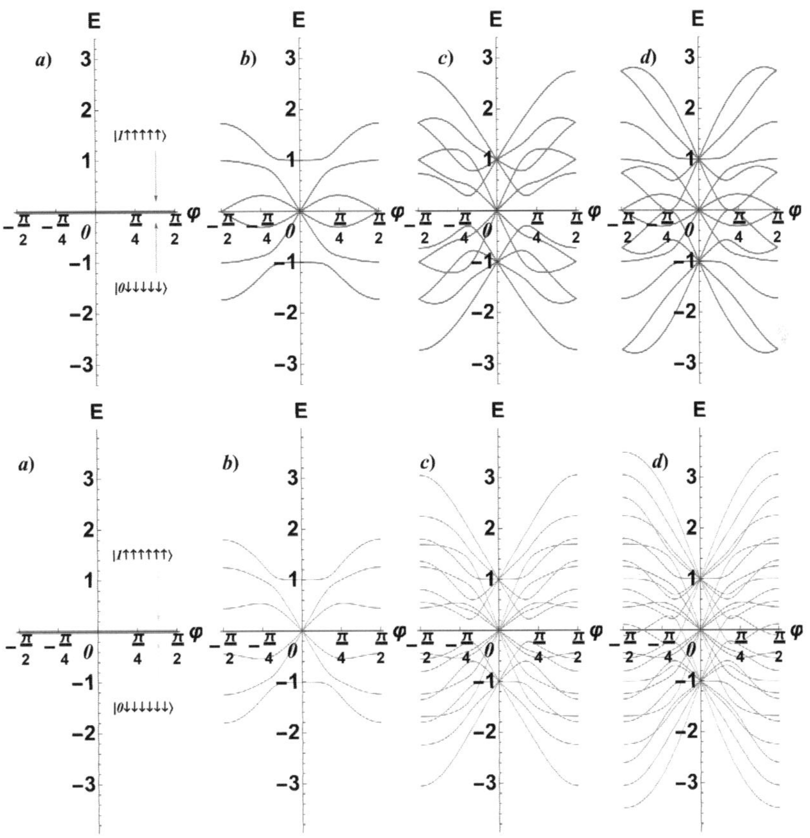

Figure 2. (color online) Energy levels sorted by the invariant for the test case of an open spin chain with NN interaction. 5 spins (top row) or 6 spins (bottom row). Levels with $|\mathcal{M}_z|$ equal to: (a) 3 $(N = 5)$ and $7/2$ $(N = 6)$ (b) 2 $(N = 5)$ and $5/2$ $(N = 6)$ (c) 1 $(N = 5)$ and $3/2$ $(N = 6)$ 0 $(N = 5)$ and $1/2$ $(N = 6)$

4. Conclusion

In this report we have presented numerical results for one photon Jaynes-Cummings XY model for a spin ring or an open spin molecules of sies of up to 10 spin-$\frac{1}{2}$'s. The resulting energy spectra represent the mixture of the photon and the spins, while for an easier comparison with the known results, for example for the a two-level atom with a Jaynes-Cummings model, we will need to project away the photon. Some improvement of the total number of included spins may achievable by using advanced spin chain simulation packages, because the one photon (i.e., the photon filling number) approximation restricts the problem into a spin type chain. Another useful case, with filling number being in the range > 1 is a subject of future work.

Acknowledgments

AAD would like to thank prof. J. Schnack (UniBielefeld) for hospitality, where part of this work was completed, and prof. N. Tonchev (ISSP, BAS) for useful comments. This work was supported by EU FP7 INERA project grant agreement number 316309.

References

[1] Jaynes E and Cummings F 1963 *Proceedings of the IEEE* **51** 89–109
[2] Gerry C and Knight P 2004 *Introductory quantum optics* (Cambridge: Cambridge University Press)
[3] Vogel W and de Matos Filho R L 1995 *Phys. Rev.* A **52** 4214–4217
[4] You J Q and Nori F 2003 *Phys. Rev.* B **68** 064509
[5] Rempe G, Walther H and Klein N 1987 *Phys. Rev. Lett.* **58** 353–356
[6] Dodonov A V, Mizrahi S S and Dodonov V V 2006 *Phys. Rev.* A **74** 033823
[7] Batle J and Casas M 2010 *Phys. Rev.* A **82** 062101
[8] Tonchev H, Donkov A A and Chamati H 2016 *Jour. Phys.: Conf. Ser.* **682** 012032
[9] Juárez-Amaro R, Zúñiga-Segundo A and Moya-Cessa H M 2015 *Appl. Math. Inf. Sci.* **9** 299–303
[10] Landau L D and Lifshitz E M 1989 *Quantum Mechanics: Non-relativistic Theory* 3rd ed (Oxford: Pergamon Press)

Photoluminescence from 20 MeV electron beam irradiated homogeneous SiO_x and composite Si-SiO_x films

D Nesheva[1], M Šćepanović[2], M Grujić-Brojčin[2], V Dzhurkov[1], S Kaschieva[1], I Bineva[1], S N Dmitriev[3] and Z V Popović[2]

[1] Institute of Solid State Physics, Bulgarian Academy of Sciences,72 Tzarigradsko Chaussee Blvd,1784 Sofia, Bulgaria

[2] Center for Solid State Physics and New Materials, Institute of Physics, University of Belgrade, Pregrevica 118, Belgrade 11080, Serbia

[3] Joint Institute for Nuclear Research, Flerov Laboratory of Nuclear Reactions, Dubna, Moskow region 141980, Russia

E-mail: nesheva@issp.bas.bg

Abstract. The effect of 20 MeV electron irradiation on the room temperature photoluminescence from homogeneous SiO_x and composite Si-SiO_x films, containing amorphous or crystalline Si nanoparticles, is studied. Layers with $x = 1.5$ and 1.7 and thickness of 200 nm were deposited on crystalline silicon substrates by thermal evaporation of SiO in vacuum. Film annealing in an inert atmosphere at 700 °C or 1000 °C for 60 min was applied to grow amorphous or crystalline silicon nanoparticles, respectively, in a SiO_x matrix. Samples from all types of films were irradiated with 20 MeV electrons at close to room temperature and a fluence of 2.4×10^{14} el.cm^{-2}. Photoluminescence was measured under excitation with the 488 nm line of an Ar^+ laser. The electron irradiation causes a decrease of the integrated photoluminescence intensity in composite samples with initial $x = 1.7$ containing amorphous or crystalline nanoparticles and $x = 1.5$ samples with Si nanocrystals. The electron irradiation of $x = 1.5$ samples with amorphous nanoparticles slightly increases the photoluminescence intensity. The obtained results are discussed in terms of electron beam induced phase separation and Si nanoparticle size increase.

1. Introduction

Electron irradiation of solids creates defects and may cause an electronic device or system to fail. It may also cause formation of amorphous or crystalline nanosized domains by decomposition of the target material [1,2]. Therefore during the last few decades there was significant activity to explore the influence of electron irradiation on various types of semiconductor devices, in particular in metal-oxide-silicon (MOS) structures [3,4].

Photoluminescence (PL) spectroscopy has been successfully used to obtain important information about changes in the defect structure upon electron irradiation of light emitting materials [5-7]. For example, the observation that the PL intensity of irradiated TiO_2 nanocrystals was much larger as compared to the PL from non-irradiated samples, has been attributed to irradiation induced defects and particle size variation [7]. Photoluminescence, Hall and X-ray photoelectron spectroscopy measurements carried out on 0.8 MeV electron beam irradiated indium-gallium-zinc oxide films have allowed to conclude that n-type conductivity was preserved in films irradiated with a low dose up to 10^{14} electrons/cm^2 while the conductivity was converted to p-type with further dose increase [5].

These observations were attributed to strong reduction of donor-like oxygen vacancy defects as a result of formation of either oxygen interstitial or zinc vacancy acceptor defects.

Due to the modification of the energy structure afforded by charge carrier confinement, silicon nanocrystals are good candidates for photonic applications involving efficient radiation recombination. The emission from silicon nanocrystals has characteristic features: PL intensity increases and shows a "blue-shift" with decreasing crystallite size. Photoluminescence spectroscopy has been applied in a study of silicon nanocrystals' growth in amorphous silicon films by assistance of a 200 keV electron beam. Various localized states' emission was measured [8,9]. It has been

Electron irradiation is most frequently carried out in high voltage transmission electron microscopes with electrons having energy of around or less than 1 MeV [10-12] but the information on the effect of high energy electrons (> 10 MeV) on the processes of the nanoparticle growth in silicon oxide matrices is very limited [1]. Our previous Raman scattering investigations of 20 MeV electron irradiation effects on homogeneous SiO_x films (initial $x = 1.3$) and composite Si-SiO_x films ($x = 1.8$), containing amorphous silicon nanoparticles, have indicated appearance of amorphous silicon phase and some structural changes in the oxide matrix of the homogeneous SiO_x films. An electron beam stimulated decrease of the defects at the a-Si/SiO_x interface in the composite films has been assumed, as well [13].

In this work, the effect of 20 MeV electron irradiation (with a fluence of 2.4×10^{14} el$\times$cm^{-2}) on homogeneous SiO_x and composite films, containing amorphous/crystalline silicon nanoparticles (a/cSi NPs, a/cSi-SiO_x composite films), is studied by photoluminescence spectroscopy and Raman scattering measurements. Layers with initial $x = 1.5$ and 1.7 deposited on crystalline silicon substrate are investigated since they show much stronger photoluminescence at room temperature than the previously studied films with initial $x = 1.3$. The obtained results are discussed in terms of electron beam stimulation of phase separation and changes in the Si nanoparticle size.

2. Experimental details

Silicon suboxide SiO_x layers with an initial composition of $x = 1.5$ and 1.7 and thickness of 200 nm were prepared by thermal evaporation of SiO at a vacuum of 1×10^{-3} Pa on p- or n-type (100) crystalline silicon (c-Si) substrates ((7.2-10) and (4-6) $\Omega\times$cm for p or n-type Si, respectively) maintained at room temperature [14]. Prior to film deposition the silicon wafers were cleaned chemically using a standard procedure for the microelectronics industry. The deposition rate and film thickness were monitored by a calibrated quartz microbalance system. All as-deposited layers were annealed at 250 °C for 30 min in an Ar atmosphere to keep them stable at ambient conditions. This annealing procedure does not lead to formation of pure silicon phase [14]. In a part of the SiO_x samples amorphous/crystalline Si nanoparticles were grown in a SiO_x matrix with $x = 1.9$, 2.0 (for films with initial $x = 1.5$, 1.7, respectively [16]) by an additional furnace annealing at 700 / 1000 °C in an argon atmosphere for 60 min [14,15].

Both kinds of films (homogeneous SiO_x films and composite a/cSi-SiO_x films) were irradiated with 20 MeV electrons with a fluence of 2.4×10^{14} el$\times$cm^{-2}. The irradiation was carried out in Microtron MT-25 in FLNR, Joint Institute for Nuclear Research, Dubna. The beam current was $I_e \approx 5$ μA. The sample temperature was controlled during the entire irradiation process and kept close to room temperature.

Photoluminescence measurements were carried out using the 488-nm line of an Ar^+ laser, Jobin–Yvon U1000 double monochromator and a photomultiplier as detector. Micro-Raman scattering measurements were performed in the backscattering geometry by a Jobin-Yvon T64000 triple spectrometer system, equipped with a confocal microscope and a nitrogen-cooled CCD detector. Raman spectra were collected by using Verdi™ G-Series Optically Pumped Semiconductor Laser operating at 532 ± 2 nm. Both PL and Raman scattering measurements were performed at room temperature in air.

3. Results and Discussion

Journal of Physics: Conference Series **764** (2016) 012018 doi:10.1088/1742-6596/764/1/012018

Raman scattering measurements were carried out to get information about the pure silicon phase in the different kind of samples. The spectra collected on non-irradiated and irradiated homogeneous SiO_x (initial $x = 1.7$) and the corresponding composite a/cSi-SiO_x films are shown in figure 1. In all spectra the strong band situated at ~ 521 cm^{-1} is due to Raman scattering from 1TO phonons of the c-Si substrate. The low intensity feature at around 300 cm^{-1} (well seen in figure 1(a)) is also due to light scattering from the c-Si substrate. The weak band at ~ 430 cm^{-1} is normally related to scattering in amorphous SiO_2 [17]. The wide band located between 930 and 1030 cm^{-1} is assigned to multi-phonon scattering in the c-Si substrate [18,19].

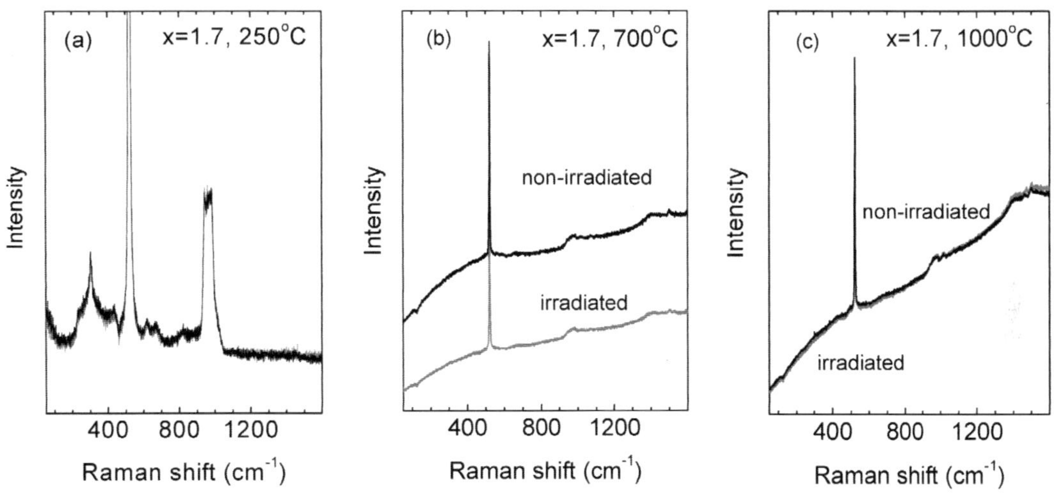

Figure 1. Raman spectra of non-irradiated and 20 MeV electron beam irradiated samples with initial x=1.7: (a) – homogeneous SiO_x film, (b) - aSi-SiO_x composite film containing amorphous Si NPs and (c) - cSi-SiO_2 composite film containing crystalline Si NPs; in (a) and (c) the spectra of non-irradiated and irradiated samples practically coincide while in (b) they have significantly different PL background. Each pair of spectra corresponds to the same y-scale but the y-scale in (a) is 40 times enlarged when compared with the y-scale in (b) and (c) to ensure better presentation of the bands observed in the Raman spectrum.

The Raman spectra of the composite films shown in figure 1(b) and (c) are strongly up-shifted with respect to the spectra of homogeneous ones (figure 1(a)) due to presence of photoluminescence background [15]. It should be noted that the Raman spectra obtained for the samples with initial $x = 1.5$ are very similar to those from figure 1, but with less intense background. Unfortunately the PL background makes difficult to draw reliable conclusions about the pure silicon phase, but it indicated that the photoluminescence spectroscopy could be applied to investigate the effect of high energy electron beam irradiation.

Photoluminescence spectra of non-irradiated and 20 MeV electron beam irradiated homogeneous and composite films with initial $x = 1.7$ and $x = 1.5$ are shown in figure 2 and figure 3, respectively. The maximum PL intensity from the $x = 1.5$ samples is more than three times lower than the intensity from the corresponding films with $x = 1.7$ and the PL from non-irradiated and irradiated homogeneous $x = 1.5$ samples is negligible. The electron beam irradiation causes some increase of the integrated PL intensity from homogeneous $x = 1.7$ film (figure 2(a)), whereas a decrease of the integrated PL intensity is observed for the aSi-SiO_x (figure 2(b)) and a/cSi-SiO (figure 2(c))

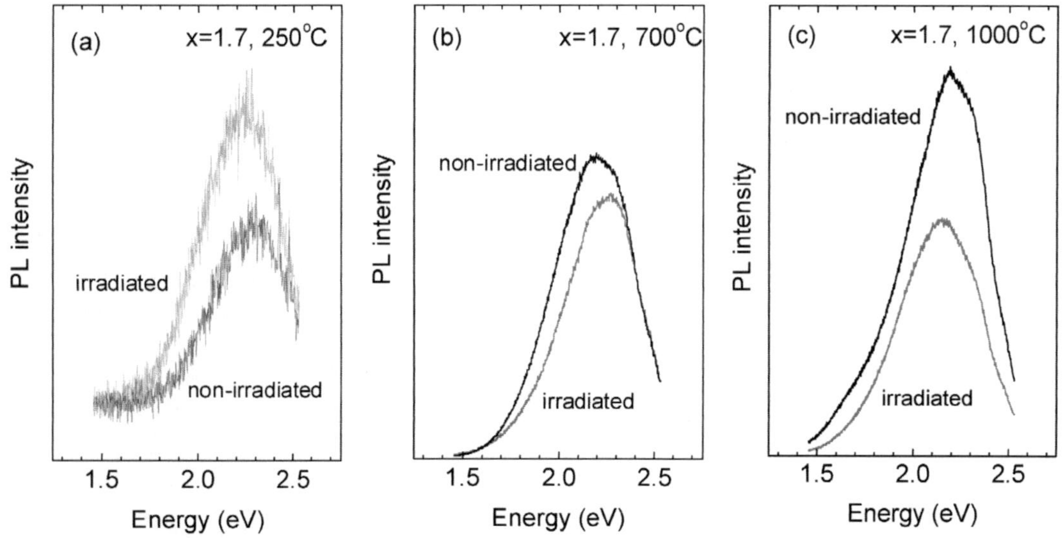

Figure 2. Photoluminescence spectra of non-irradiated and 20 MeV electron beam irradiated samples with initial $x = 1.7$: (a) – homogeneous SiO_x films, (b) - aSi-SiO_x composite films containing amorphous Si NPs and (c) - cSi-SiO_2 composite films containing crystalline Si NPs. Each pair of spectra corresponds to the same y-scale but the y-scale in (a) is 4 times enlarged when compared with the same y-scale in (b) and (c) in order to ensure better presentation of the PL bands.

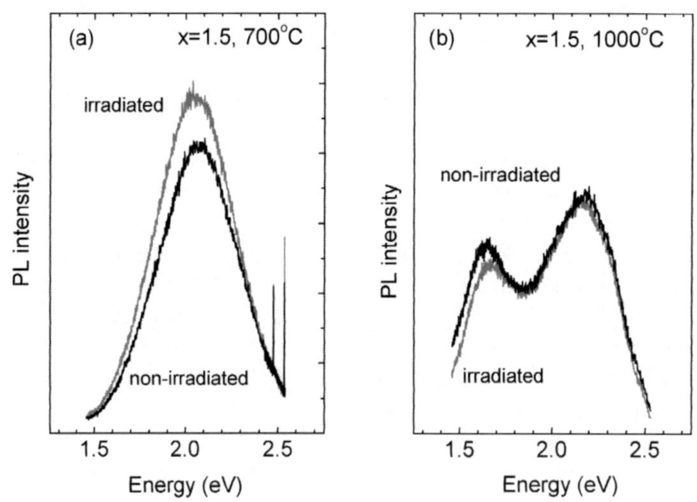

Figure 3. Photoluminescence spectra of non-irradiated and 20 MeV electron beam irradiated samples with initial $x = 1.5$: (a) - aSi-SiO_x composite films containing amorphous Si NPs and (b) - cSi-SiO_2 composite films containing crystalline Si NPs. All spectra correspond to the same y-scale.

composite films, being stronger for the films with crystalline Si nanoparticles. In the $x = 1.5$ samples the electron beam induced changes are weak; in the samples with aSi NPs the electron irradiation causes slight PL intensity increase (figure 3(a)), whereas the spectra of non-irradiated and irradiated cSi-SiO_2 composite films with nanocrystals are quite similar (figure 3(b)).

Our previous investigations on non-irradiated samples have shown [15,20] that when exciting with the 488 nm line, the PL spectrum consists of two bands: a band related to radiative recombination via defects in the oxide or at the Si/SiO_x interface (band 1 peaked in the range 2.2-2.3 eV) and a second band related to radiative recombination of electrons and holes localized in Si nanoparticles (band 2). Normally band 1 is related to radiative recombination in non-bridging oxygen hole center (NBOHC, PL centered at ~2 eV) and/or positively charged oxygen vacancy (E' center, PL centered at ~ 2.3 eV) [21,22]. For the aSi-SiO_x films band 2 (i.e. PL from Si NPs) has maximum in the 'orange-red' spectral range which moves from ~ 2.1 to ~ 1.7 eV [15, 20] with changing NP size while for the cSi-SiO_x the band 2 maximum moves from ~ 2.0 to ~ 1.6 eV [15]. It has been shown that the intensity of PL from Si NPs strongly decreases with increasing nanoparticle size [15,20]. Keeping in mind all those results, the PL spectra of both non-irradiated and electron irradiated samples have been deconvoluted into two Gaussian bands and the results obtained for the composite films with initial $x = 1.7$ are shown in figure 4. As seen from the figure, in both types of composite films the intensity of the PL from NPs (the band at ~ 2.15 eV (compare figure 4(a) and (b)) and the band at ~ 2.0 eV (compare figure 4(c) and (d)) decreases upon electron irradiation. The result is similar for the cSi-SiO_2 $x = 1.5$ films (figure 3 (b)). Only in the aSi-SiO_x films a slight increase of the band 2 intensity is observed (figure 3 (a)).

The electron beam irradiation could cause: (i) phase separation or rearrangement of the defect structure in the homogeneous films; (ii) further phase separation in aSi-SiO_x composite films with keeping NP size constant and only increasing the amount of the pure aSi phase; (iii) NP size increase in composite films which results in an intensity decrease and some red shift; (iv) improvement of the interface between the nanoparticles and oxide matrix, as assumed in Ref. [13]. The mechanism (ii) is not expected in composite films with cSi NPs since a complete phase separation takes place upon 60 min annealing at 1000°C [14]. The mechanisms (ii) and (iv) should cause a PL intensity increase. Hence in all films with the irradiation induced PL intensity reduction most likely the electron beam irradiation causes some nanoparticle size increase (mechanism (iii)). This could be easy understood since the matrix composition in all those films is $x \sim 2.0$ [14, 16]. Only in the aSi-SiO_x films with initial $x = 1.5$, in which the oxygen matrix content is $x < 2.0$, further phase separation (mechanism (ii)) could take place and could have more pronounced effect than the effect of the nanoparticle growth, thus causing the observed slight PL intensity increase. We notice that the Raman scattering increase observed in aSi-SiO_x films with initial $x = 1.3$ [13] can be also due to electron beam induced phase separation (at least partially, together with the annealing effect). Finally, the PL increase found for the $x = 1.7$ homogenenous films (figure 2 (a)) could be related to electron induced formation of few very small aSi clusters but it could be also due to some increase of the NBOHC or E' defects.

Upon electron irradiation the intensity of the PL from the oxide matrix (band 1) shows an increase in the aSi-SiO_x films (figure 4 (a), (b)) and a decrease in the cSi-SiO_2 films (figure 4 (c), (d)). However in the $x = 1.5$ samples (the fit results are not shown) the intensity of band 1 strongly decreases in the aSi-SiO_x films and does not change cSi-SiO_2 films. Hence, no clear tendency is observed with changing the film composition and annealing temperature. Probably both the initial composition and the annealing temperature considerably affect the density of the defects in the oxide matrix and the observed various PL intensity changes are result of a specific contribution of both effects. Because of this the band 1 behaviour was not helpful for clarifying the origin of the PL from irradiated $x = 1.7$ homogeneous films.

4. Conclusions

Homogeneous 200 nm thick SiO_x films ($x = 1.5$ and 1.7) on c-Si substrate and composite a/cSi-SiO_x films containing amorphous/crystalline Si nanoparticles have been irradiated with 20 MeV electrons

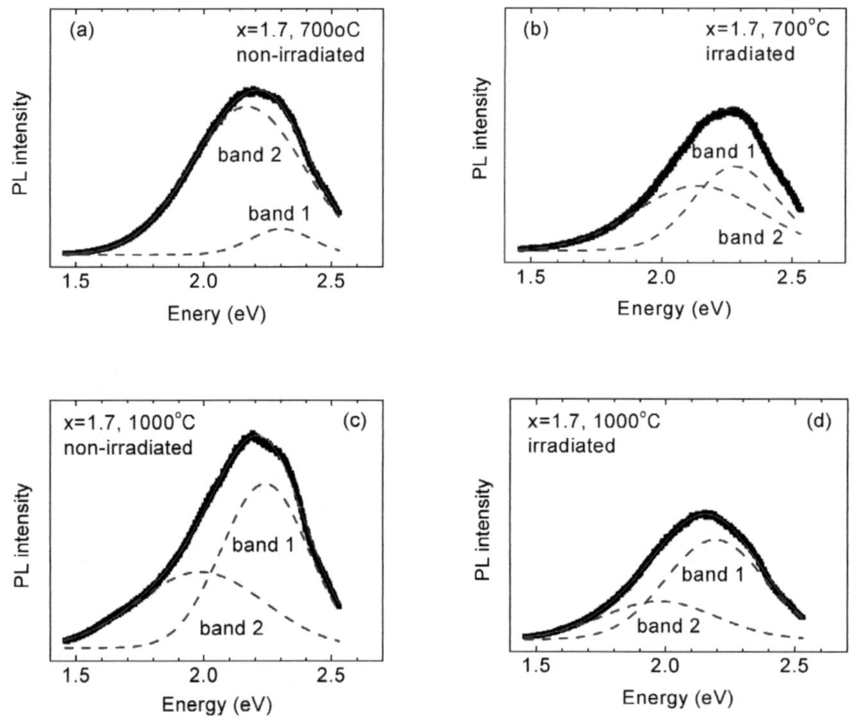

Figure 4. Deconvolution of the photoluminescence spectra of 20 MeV electron beam irradiated composite samples with initial $x = 1.7$: (a), (b) – non-irradiated and irradiated aSi-SiO_x films, respectively, and (c), (d) - non-irradiated and irradiated cSi-SiO_2 films, respectively. All spectra correspond to the same y-scale.

at a fluence of 2.4×10^{14} el.cm^{-2}. The results from photoluminescence measurements under excitation with the 488 nm of an Ar^+ laser have revealed an electron beam induced intensity increase of PL from the $x = 1.7$ homogenious samples. The PL increase has been related to appearance of small amount of aSi nanoparticles due to electron beam induced initial phase separation and/or an increase of the NBOHC/E' defect density. An increase of the PL from the $x = 1.5$ composite films containing aSi NPs has been found and ascribed to increase of the total amount of the pure aSi phase due to further phase separation stimulated by the electron beam irradiation in samples with under stoichiometric oxide matrix. The observed intensity decrease of the PL from composite films with nearly stoichiometric oxide matrix (initial $x = 1.7$ containing a/cSi NPs and composite films with initial $x = 1.5$ containing cSi NPs) has been attributed to electron beam induced increase of nanoparticle size.

Acknowledgements
This work is partially supported under Grant № 22/15.01.2016/N16 from JINR, Dubna, Russia, a bilateral grant supported by the Bulgarian Academy of Sciences and Serbian Academy of Sciences and Arts, and Project III45018 of the Ministry of Education, Science, and Technological Development of Republic of Serbia. Financial support from the Project INERA under the contract REGPOT 316309 for the paper presentation at the INERA Conference 2016 "Vapor Phase Technologies for Metal Oxide and Carbon Nanostructures" is gratefully acknowledged.

Journal of Physics: Conference Series **764** (2016) 012018 doi:10.1088/1742-6596/764/1/012018

References

[1] Kaschieva S and Dmitriev S N 2010 *Radiation Defects in Ion Implanted and/of High-Energy Irradiated MOS Structures* (New York: Nova Science Publishers)
[2] Krasheninnikov V and Nordlund K 2010 *J. Appl. Phys.* **107** 071301 and references therein
[3] Fourches N T and Nenoi M 2012 *Current Topics in Ionizing Radiation Research* **32** 741 (Rijeka: InTech)
[4] Oldham T R 2003, *IEEE Trans. Nuclear Sci.* **50** 483
[5] Jung S H, Moon H J, Ryu M K, Cho K I, Bae B S and Yun E -J 2012 *Journal of Ceramic Processing Research* **13** Special. 2 s246
[6] Egilsson T, Henry A, Ivanov I G, Lindström J L and Janzén E 1999 *Phys. Rev.* **B 59**, 8008
[7] Priyanka K P, Sunny J, Anu T S, Jaseentha O P and Varghese T 2012 *International Journal of Emerging Technology and Advanced Engineering* **2** 130
[8] Huang W-Q, Liu S-R, Huang Z-M, Dong T-G, Wang G and Qin C-J 2015 *Sci. Rep.* **5** 9932
[9] Huang W-Q, Liu S-R, Huang Z-M, Dong T-G, Wang G and Qin C-J 2015 *Sci. Rep.* **5** 16682
[10] Takeguchi M, Tanaka M and K Furuya 1999 *Appl. Surf. Sci.* **146** 257
[11] Chen G S, Boothroyd C B and Humphreys C J 1998 *Phil. Mag.***A78** 491
[12] Du X-W, Takeguchi M, Tanaka M and Furuya K 2003 *Appl. Phys. Lett.* **82** 1108
[13] Nesheva D, Dzhurkov V, Šćepanović M, Bineva I, Manolov E, Kaschieva S, Nedev N, Dmitriev S N and Popović Z V 2016 J. Phys.: Conf. Series **682** 012012
[14] Nesheva D, Bineva I, Levi Z, Aneva Z, Merdzhanova T and Pivin J C 2003 *Vacuum* **68** 1
[15] Nesheva D, Raptis C, Perakis A, Bineva I, Aneva Z, Levi Z, Alexandrova S and Hofmeister H 2002 *J. Appl. Phys.* **92** 4678
[16] Donchev V, Nesheva D, Todorova D, Germanova K and Valcheva E 2012 *Thin Solid Films* **520** 2085
[17] Rahmani A, Benoit M and Benoit C 2003 *Phys. Rev.* B **68** 184202
[18] Temple P A and Hathaway C E 1973 *Phys. Rev.* B **7** 3685
[19] Borowicz P, Latek M, Rzodkiewicz W, Łaszcz A, Czerwinski A and Ratajczak J 2012 *Adv. Nat. Sci.: Nanosci. Nanotechnol.* **3** 045003
[20] Bineva I, Nesheva D, Aneva Z and Z. Levi Z 2007 **126**, 497
[21] Vázquez-Valerdi D E, Luna-López J A, Carrillo-López J, García-Salgado G, Benítez-Lara A and Espinosa-Torres N D 2014 *Nanoscale Research Letters* **9** 422
[22] Lin C-J and Lin G-R 2005 *IEEE J. Quantum Electronics* **41** 441

Characterization of nanostructured TiO_2:Ag films: structural and optical properties

T Ivanova[1], A Harizanova[1], T Koutzarova[2] and B Vertruyen[3]

[1]*Central Laboratory of Solar Energy and New Energy Sources, Bulgarian Academy of Sciences, blvd. Tzarigradsko chaussee 72, Sofia, Bulgaria*
[2]*Institute of Electronics, Bulgarian Academy of Sciences, blvd. Tzarigradsko chaussee 72, Sofia, Bulgaria*
[3]*LCIS/SUPRATECS, Institute of Chemistry B6, University of Liege, Sart-Tilman, B-4000 Liege, Belgium*

E-mail: tativan@phys.bas.bg

Abstract. TiO_2:Ag nanocomposites have been prepared by sol-gel method with varying silver concentration. Different technological approaches are employed to study the formation of Ag nanoparticles in titanium dioxide matrix. The obtained thin films are either thermally treated at temperatures from 300 to 600°C or UV irradiated for 15 minutes between layer deposition. XRD and FTIR studies reveal that Ag is incorporated in TiO_2 films as nanoparticles and no Ag oxide phases are detected. Optical characterization performed by UV-VIS spectroscopy confirms the formation of silver nanoparticles. The influence of thermal treatment and UV radiation on the optical and structural properties is studied.

1. Introduction

Titanium oxide is one of the most widely studied metal oxide semiconductors for various applications due to its interesting chemical, electrical and optical properties, chemical stability, non toxicity, high transparency and high refractive index, cheapness and commercial availability, antibacterial and hydrophilic ability [1]. TiO_2 has many uses such as solar energy conversion, photocatalyst, photoelectrochemical splitting of water, purification and treatment of polluted water and air and can be found in multitude of applications: gas sensors, photochromic, electrochromic and optoelectronics devices, ceramics [2]. Titanium oxide can exist in different titania polymorphs such as rutile, anatase and brookite. These three polymorphs are structurally similar: the Ti atom is octahedrally coordinated, but differs in the way the TiO_6 octahedra are linked within anatase, rutile and brookite. The anatase tetragonal crystal structure is made of distorted TiO_6 octahedral sites sharing four corners. The brookite (orthorhombic, the structure with the lowest symmetry among the polymorphs) is also made of distorted octahedral sites elongated along the c-axis, and each octahedron shares three edges with adjacent octahedra. The rutile structure (tetragonal, thermodynamically stable phase) is made of chains of TiO_6 octahedra that share a vertex along the c-axis. The anatase transforms into rutile upon heating, exothermally and irreversibly over a certain range of temperatures [3]. The brookite phase is the rarest of the TiO_2 polymorphs. Anatase TiO_2 has a band gap of 3.2 eV (385 nm), while the rutile phase has a smaller band gap of 3.0 eV (410 nm) [4].

Ag presents numerous interesting properties, such as high catalityc efficiency, low cost compared to other noble metals, and high oxygen adsorption reactivity. TiO_2 thin films doped with Ag have shown many interesting characteristics. Moreover, Ag metallic ion is known as a high mobility cation, which has been used in resistive switching devices [5]. In this sense, TiO_2 thin films doped with Ag NPs are attractive material for applications in sensors, photocatalysis, as plasmonic colors, hot spots, thermal phototherapy, plasmon assisted solar energy conversion, information storage devices, displays, smart windows and switches. Silver nanoparticles embedded in dielectric films exhibit optical responses arising from the excitation of surface plasmons in the metal nanoparticles (NPs) that induces optical absorption. The frequency and shape of the surface plasmon resonance (SPR) depends on the size, shape and volume fraction of Ag NPs.

In this work, TiO_2:Ag nanocomposites are prepared by sol-gel method with varying silver concentration. Different technological approaches are employed to study the formation of Ag nanoparticles in titania matrix. The obtained thin films are either thermally treated at temperatures from 300 to 600°C or UV irradiated for 15 minutes.

2. Experimental

The details of sol solution preparation for TiO_2 have been previously reported [6]. It must be noted that the sol solution is proven to be stable more than 2 years, maintaining its excellent film forming properties. The Ag precursor used was silver nitrate as it was added 0.1, 0.5 and 1% wt $AgNO_3$ in titanium sol solution. The thin films were deposited by spin-coating procedure at 8000 rpm. For the first group of samples, the spin coated procedure was followed by low temperature firing (150°C/30 min) and repeating the layer deposition 5 times in order to increase the film thickness. For the second group, UV treatment for 15 min. was applied between layers. The TiO_2 and TiO_2:Ag films have been thermally treated at 300, 400, 500 and 600°C in oxygen for 1 hour. The film thickness was in the range of 240 to 200 nm depending on the annealing temperatures. The substrates used were Si wafers and quartz. XRD spectra of the films were recorded by means of XRD diffractometer Bruker D8, at the grazing angle 2°, step time of 8 s and step of 0.1°. FTIR measurements were performed in the spectral region 350-1200 cm^{-1} by Shimadzu FTIR Spectrophotometer IRPrestige-21. Optical measurements were performed using UV-3600 Shimadzu spectrophotometer.

3. Results and discussions

Figure 1 present XRD spectra of TiO_2:Ag films, obtained by sol B (0.5 wt% Ag precursor) annealed at temperatures from 300 to 600°C. The lowest annealing temperature results in amorphous film structure and increasing the annealing temperature induces crystallization.

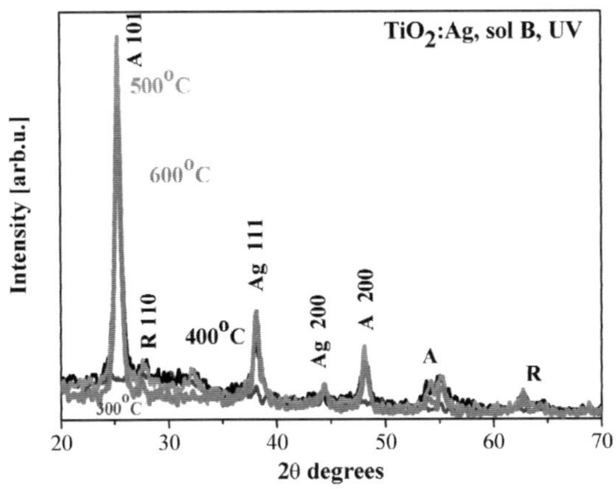

Figure 1. XRD spectra of sol-gel TiO_2:Ag films, obtained by sol B (0.5 wt% Ag precursor) annealed at temperatures from 300 to 600°C. The marks A and R denote anatase and rutile phase, respectively.

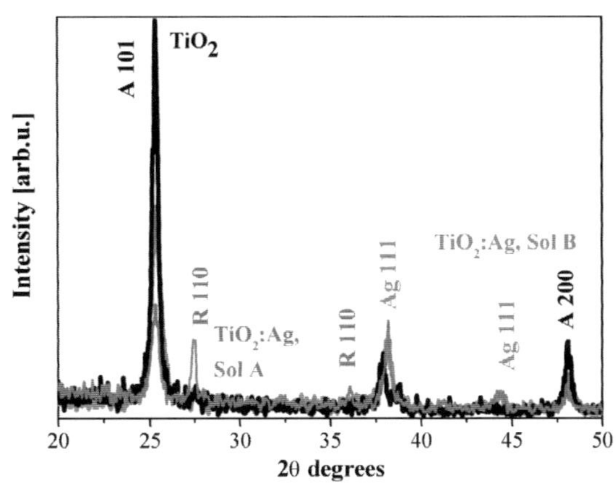

Figure 2. Comparison of XRD spectra of pure TiO_2 and TiO_2:Ag films, obtained by sol A (1 wt% $AgNO_3$) and sol B (0.5 wt% $AgNO_3$). The TiO_2:Ag films were UV treated between layers. The samples are finally annealed at 600°C.

The two crystal phases – anatase (PDF card №00-004-0477), and rutile (JCPDS №21-1276) are clearly presented in the XRD spectra of 400 to 600°C annealed films. The previous study of sol-gel undoped TiO_2 [6] films showed that the rutile phase begins to form after high temperature annealing (at and above 800°C). The spectra manifest a diffraction reflection due to presence of Ag (cubic structure, PDF card №00-004-0783), indicative that silver is incorporated in the film structure as metallic nanoparticles. For annealed at 600°C film, two strong lines Ag 111 and Ag 200 attributed to metallic Ag are detected. Similar results had been previously reported for XRD study of thermally treated TiO_2:Ag films, obtained by sol A (1 wt% Ag precursor) [7]. Figure 2 shows the comparison of XRD spectra of TiO_2:Ag films with different silver incorporation. It is clearly seen that higher Ag content results in better crystallinity with stronger lines related to rutile TiO_2 and metallic cubic Ag. The spectrum of undoped TiO_2 film reveals that the film is fully crystallized in anatase phase (see figure 2, black line).

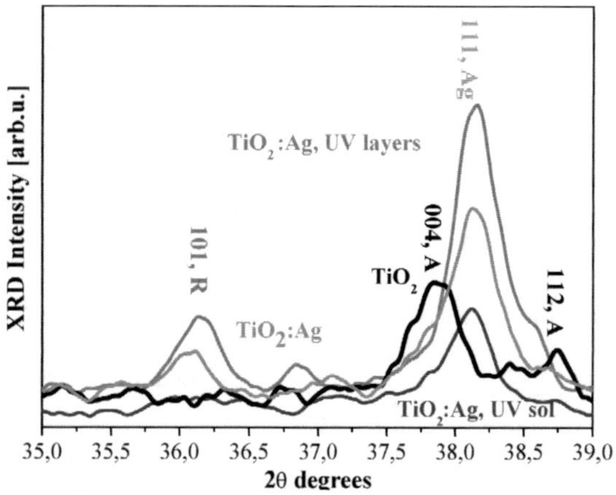

Figure 3. XRD spectra of sol-gel TiO_2 and TiO_2:Ag films. TiO_2:Ag films are obtained by sol A (1 wt% Ag precursor) in three types: thermal treatment (150°C/10 min) between layers, UV treatment of sol before film deposition and UV treatment between layers. The samples are finally annealed at 600°C.

Figure 3 shows XRD spectra of undoped TiO_2 and TiO_2:Ag films, obtained by sol A (1 wt% Ag precursor), undergone three different treatments: thermal treatment (150°C/10 min) between layers, UV treatment of sol before the film deposition and UV treatment between layers. All the samples are finally annealed at 600°C. The films with incorporated Ag nanoparticles present a fraction of rutile titania and metallic silver. The UV treatment of films leads to strongest XRD lines of silver. The crystallite sizes have been determined for the existing crystal phases in film structure (see table 1). It is very interesting that the biggest Ag crystallites are found out for TiO_2:Ag from sol A, treated at 600°C. For UV treated films, it is determined that silver nanoparticles grew bigger, when the sol has been preliminary UV radiated. The crystallite sizes for anatase and rutile phase are also determined.

Table 1. Estimated crystallite sizes of TiO_2:Ag films.

	TiO_2	TiO_2:Ag sol A, thermally treated films	TiO_2:Ag sol B, thermally treated films	TiO_2:Ag sol A, UV treated films	TiO_2:Ag sol A, UV treated sol
Anatase	12.4 nm	10.9 nm	17.7 nm	8.2 nm	12.1 nm
Rutile		18.4 nm	8.8 nm	25.1	15.4 nm
Ag		80.2 nm	8.3 nm	19.7 nm	25.0 nm

XRD analysis reveals diffraction features characteristic for metallic state of silver with no Ag-O phases detected. Another conclusion is that Ag NPs induce anatase-rutile phase transformation at lower annealing temperatures than it appeares for undoped TiO_2 films. UV treatment results in better expressed rutile phase compared to only thermally treated films. The lower Ag concentrations the smaller Ag NPs crystallites are grown.

FTIR study has been performed in details for thermally and UV treated films, but there is presented only the spectra of sol-gel TiO_2:Ag films, obtained by sol A and sol B. The Ag doped films have been either annealed at 150°C or UV treated between layers. The samples are finally annealed at 600°C (see figure 4). The weak IR band at 1058 cm^{-1} is assigned to Si-O bonding (due to Si substrate).

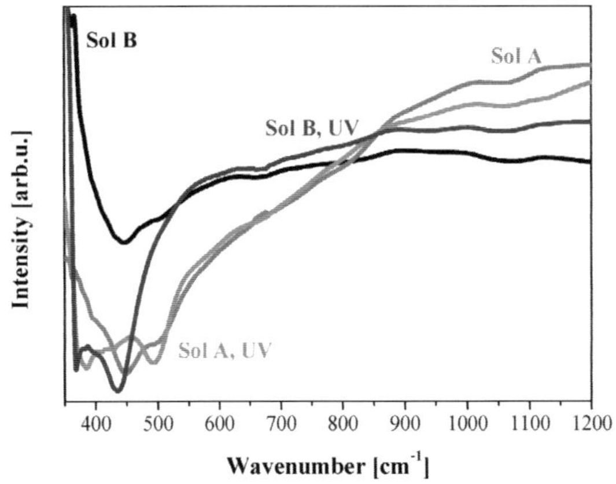

Figure 4. FTIR spectra of sol-gel TiO_2:Ag films, obtained by sol A (1 wt% Ag precursor) and sol B (0.5wt% Ag precursor). The Ag doped films have been either annealed at 150°C or UV treated between layers. The samples are finally annealed at 600°C.

A weak band at 668 cm^{-1} can be observed in all spectra, which some authors assigned to integrated anatase and rutile structure [8]. The FTIR spectra of thermally treated films show a broad and strong absorption band near to 445 cm^{-1} and a weak band at 504 cm^{-1}. UV treatmeant results in splitting of the strong band in two separate lines at 385 and 430 cm^{-1} (UV, Sol A) and at 370 and 435 cm^{-1} (UV, Sol B). Similar behaviour was previuosly detected [9] and it is caused by the formation of rutile fraction. The bands located in the range 430 -435 cm^{-1} are attributed to Ti-O bonds of crystalline anatase titania, on the other hand the 445 cm^{-1} is assigned to rutile phase. The band at 385 cm^{-1} is also referred to rutile phase. No lines, assigned to Ag-O bonds are detected.

FTIR analysis reveals that the silver incorporation reflects on shapes and intensity of absorption bands with no traces of Ag-O bonds. Characteristic absorption bands assigned to rutile titanium dioxide are observed for samples annealed at 600°C and for TiO_2:Ag films after UV radiation. The structure of films consists of mixture of antase and rutile phase and UV treatment leads to well prounnced absorption bands due to rutile TiO_2 in comparison with thermally treated films. FTIR investigation confirms the conclusions derived from XRD analysis that no Ag-O bonding is observed and the silver incorporation induces anatase-rutile transformation.

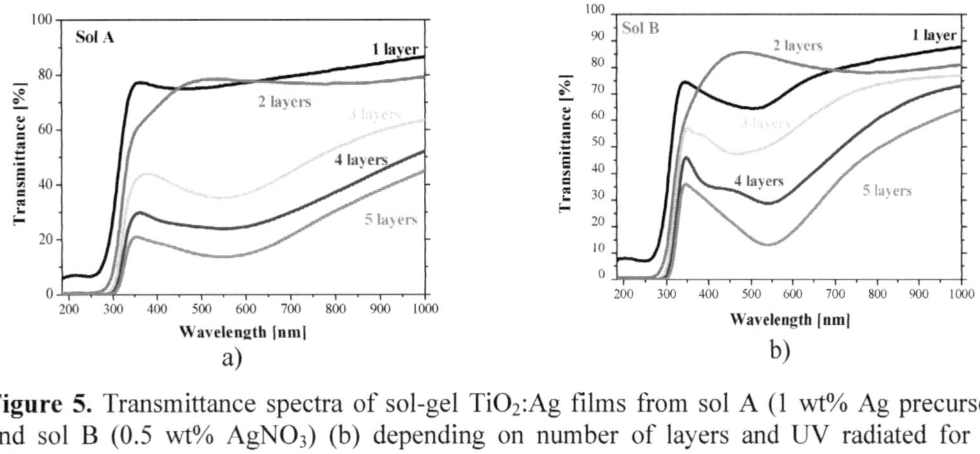

Figure 5. Transmittance spectra of sol-gel TiO_2:Ag films from sol A (1 wt% Ag precursor) and sol B (0.5 wt% $AgNO_3$) (b) depending on number of layers and UV radiated for 15 minutes between layers.

For optical characterization, the thin films were deposited on quartz substrates. Optical properties of nanocomposite layers of TiO_2:Ag films are studied by transmittance measurements as a function of number of layers, silver concentrations, thermal and UV treatment.

Figure 5 shows the transmittance spectra of sol-gel UV treated TiO_2:Ag films from sol A (1 wt% Ag precursor) and sol B (0.5 wt% $AgNO_3$) with varying number of layers. As can be observed the treansparency is decreased with increasing the layers and the shape of the absorption bands is changed. The samples from sol A manifest very broad bands in the spectral range of 510-550 nm, which can be due to the plasmonic features induced by larger Ag nanoparticles or/and overlapping of higher order plasmon modes [10]. It is very interesting to observe that better expressed surface plasmon resonance (SPR) bands are observed for UV treated TiO_2:Ag films from sol B, the lower silver concentration. The one layered film possesses a strong band at 506 nm. For three layered film the plasmonic bands are located at 363 and 470 nm, where the 363 nm feature can be related to the quasispherical Ag nanoparticles or Ag oligomers with different sizes [11] and the other band is also connected with SPR of near spherical size Ag NPs. The four layered film manifest two absorptions at 400 and at 540 nm and the thickest film has a very deep feature at 540 nm. The SPR at 400 nm is characteristical sign for presence of spherical Ag NPs [12]. The manifestation of the SPR bands is a proof for the formation of silver nanoparticles and the broadness of these bands is also indicative for Ag NPs with broad size distribution and various shapes [12].

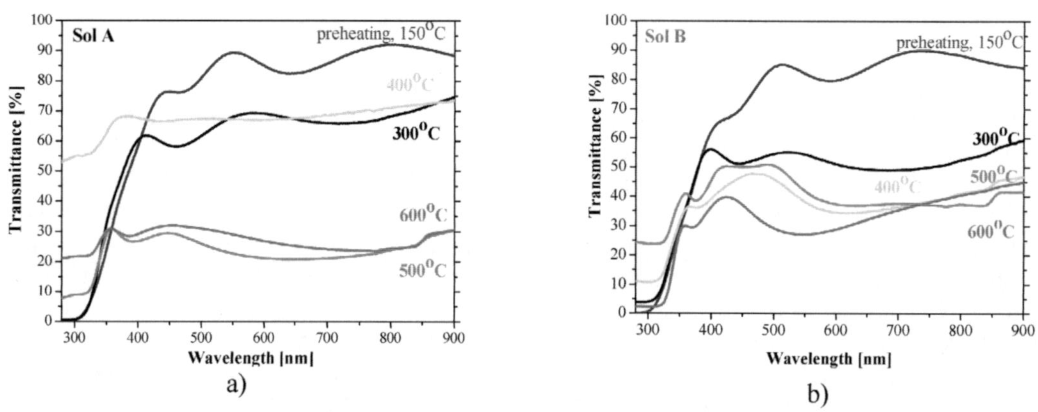

Figure 6. Transmittance spectra of sol-gel TiO_2:Ag films (5 layers) from sol A and sol B, thermally treated at different temperatures.

Figure 6 presents the optical spectra of TiO_2:Ag films (5 layers) from sol A and sol B, thermally treated at different temperatures. All the films show plasmonic absorption but the position and intensity depends significantly on the annealing temperatures. For all samples a band near 400 nm can be observed – an SPR due to spherical Ag NPs but there are also broader bands at higher wavelengths which can be related to bigger nanoparticles, to non spherical shapes and to higher plasmonic modes.

Comparison of the optical absorption of low temperature annealed undoped TiO_2, TiO_2:Ag films (150°C, preheating temperature) and only UV treated films is given in figure 7. The low absorption of undoped TiO_2 indicates its high transparency. The only thermally treated films with Ag NPs also reveal relatively low absorption and the UV treatment drastically increases the absorption in the visible spectral range. The transmittance studies prove that Ag is embedded in TiO_2 films as nanoparticles as plasmon absorptions are clearly exhibited in the visible spectral range. Absorption peaks, a sign for spherical Ag nanoparticles are observed along with broader absorption bands that can be assigned to bigger nanoparticles or to different more complex shapes of Ag NPs. Plasmonic nanostructured thin films with increased absorption can be useful in thin-film solar cells through efficient scattering of the incident light in semiconductor absorber [13].

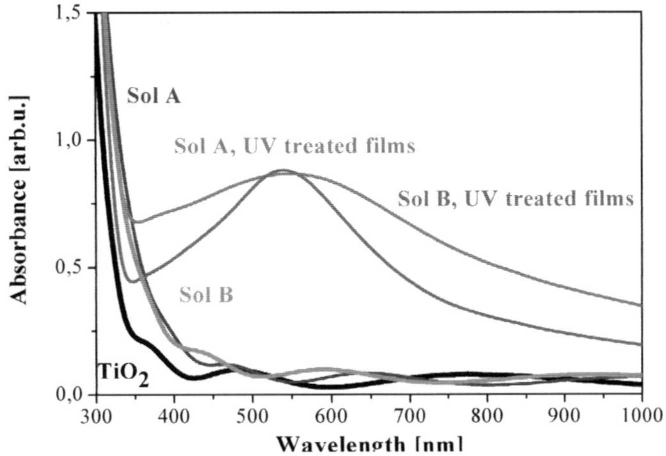

Figure 7. Optical absorbance of sol-gel TiO_2 and TiO_2:Ag films (5 layers) from sol A and sol B, preheated at 150°C or UV treatment.

From transmittance measurements it can be estimated the optical band gap (see figure8). The estimated optical band gaps (E_g) for TiO_2 films are higher than those reported for bulk anatase TiO_2 - 3.2 eV and bulk rutile 3.03 eV [7]. The widening of the band gap is due to the nanocrystalline nature of the films, grain size, variation of the crystal structure.

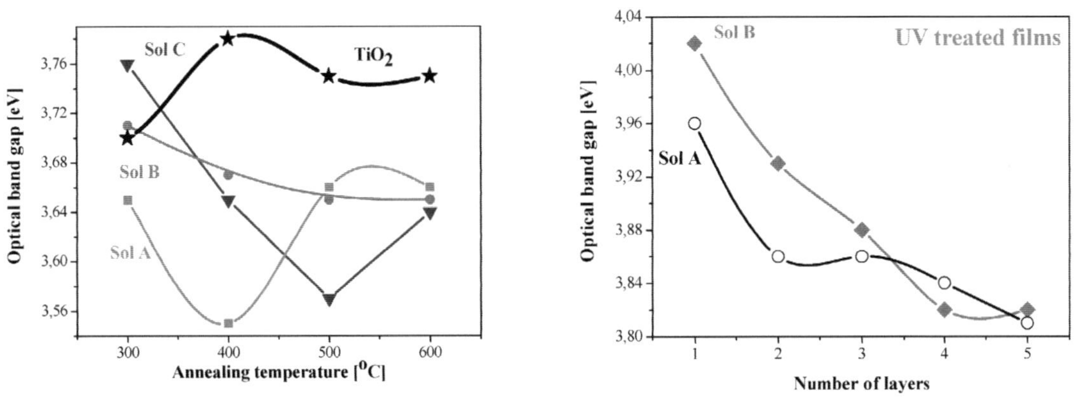

Figure 8. Optical band gap of sol-gel TiO_2:Ag films from sol A (1 wt% $AgNO_3$), B (0.5 wt% Ag NO_3)and C (0.1 wt% $AgNO_3$) (5 layers) thermally treated at different temperatures compared to undoped TiO_2 and optical band gap values of UV treated TiO_2:Ag films from sol A and B depending on number of layers.

The optical band gap of thermally treated TiO_2:Ag films are smaller than those of pure titania due to effect of Ag nanoparticles. The observed decrease of E_g with silver addition is reported by other authors [14]. The UV radiated films show very different optical behavior. They possess optical band gaps ranging from 3.80 to 4.04 eV depending on the number of layers and silver concentration. Many authors reported a relation between the optical band gap and crystallites sizes, which can be related to quantum confinement [16], the presence of the two crystal phases of TiO_2 and Ag nanoparticles.

4. Conclusions

The sol-gel approach offers a facile deposition method of Ag nanoparticles TiO_2 films. The results indicate that the silver incorporation and its concentration affect the crystallinity and the optical properties of TiO_2:Ag thin films. The annealing temperatures and UV treatment influence greatly on the structural and optical behaviour of TiO_2:Ag nanostructured thin films. XRD analysis reveals that the silver is crystallized in metallic state. It has been found that Ag presence facilitates the anatase to rutile transformation and the UV treatment results in better expressed rutile phase compared to only thermally treated films. FTIR investigation show no absorption bands related to Ag-O bondings and IR lines characteristic for anatase and rutile phase of TiO_2 are observed. Optical characterization reveals that Ag nanoparticles are successfully formed in titanium dioxide matrix as surface plasmon resonance effect is detected. Optical absorption in the visible spectral range exhibits plasmonic features characteristic for spherical Ag nanoparticles and broader absorption bands that can be assigned to bigger nanoparticles or higher plasmonic modes.

Acknowledgements

Financial support from the project INERA under the contract REGPOT 316309 for the paper presentation at the INERA Conference 2016 “Vapor Phase Technologies for Metal Oxide and Carbon Nanostructures” is gratefully acknowledged

References

[1] Wang X, Hou X, Luan W, Li D and Yao K 2012 *Appl. Surf. Sci.* **258** 8241– 8246
[2] Bensouici F, Souier T, Dakhel A, Iratni A, Tala-Ighil R and Bououdin M 2015 *Superlatt. Microstr.* **85** 255–265
[3] Tobaldi D M, Pullar R C, Gualtieri A F, Seabra M P and Labrinch J A 2013 *Acta Mater.* **61** 5571–5585
[4] Paul S and Choudhury A 2014 *Appl. Nanosci.* **4** 839–847
[5] Sheini N A and Rohani1 M 2016 *IOP Conf. Series: Mater. Sci. Engin.* **108** 012033
[6] Harizanov O and Harizanova A 2000 *Solar Energy Mater. Solar Cells* **63** 185-195
[7] Ivanova T, Harizanova A, Koutzarova T and Vertruyen B 2013 *Opt. Mater.* **36** 207-213
[8] Wang W, Chen J, Gao M, Huang Y-X, Zhang X and Yu H-Q 2016 *Appl. Catal. B* **195** 69-76
[9] Vitanov P, Harizanova A and Ivanova T 2012 *IOP: Confer. Series* **356** 012041
[10] Shan G, Zheng Zh, Chen Sh and Liu Y 2012 *Colloid Surf. B* **94** 157-162
[11] Akgun B A, Durucan C, Mellot N P 2011 *J. Sol. Gel Sci. Techn.* **58** 277
[12] Velgosova O, Mraznikova A and Marcincakova R 2016 *Mater. Lett.* **180** 336-339
[13] White T P and Catchpole K R 2012 *Appl. Phys. Lett.* **101** 073905
[14] Zhao C, Krall A, Zhao H, Zhang Q and Li J 2012 *Inter. J. Hydrogen Energy* **37** 9967-9976
[15] Wei Ch –H and Chang Ch-M 2011 *Mater. Transactions* **52** 554 - 559
[16] Djaoued Y, Ozga K, Wojciechowski A, Reshak A H, Robichaud J and Kityk I V 2010 *J. Alloys Compounds* **508** 599–605

Journal of Physics: Conference Series **764** (2016) 012020 doi:10.1088/1742-6596/764/1/012020

Bioactive glass thin films synthesized by advanced pulsed laser techniques

N Mihailescu[1], George E Stan[2], C Ristoscu[1], M Sopronyi[1] and Ion N Mihailescu[1*]
[1] National Institute for Lasers, Plasma and Radiation Physics, 409 Atomistilor Street, Magurele, Ilfov, RO-77125, Romania;
[2] National Institute of Materials Physics, 405A Atomistilor Street, Magurele-Ilfov, RO- 077125, Romania

E-mail: ion.mihailescu@inflpr.ro

Abstract. Bioactive materials play an increasingly important role in the biomaterials industry, and are extensively used in a range of applications, including biodegradable metallic implants. We report on Bioactive Glasses (BG) films deposition by pulsed laser techniques onto biodegradable substrates. The BG coatings were obtained using a KrF* excimer laser source (λ= 248 nm, $\tau_{FWHM} \leq 25$ ns). Their thickness has been determined by Profilometry measurements, whilst their morphology has been analysed by Scanning Electron Microscopy (SEM). The obtained coatings fairly preserved the targets composition and structure, as revealed by Energy Dispersive X-Ray Spectroscopy, Grazing Incidence X-Ray Diffraction, and Fourier Transform Infra-Red Spectroscopy analyses.

1. Introduction

Bone fractures caused by various unwanted accidents or different diseases heal at different times, depending on the type of fracture and patients age. Certain fractures in young kids can heal as quickly as 3 weeks, while it may take as long as 6 weeks for the same kind of fracture to heal in teenagers. And some fractures can take more than 10 weeks to mend [1]. It is expected that due to the increase of life expectancy and access to quality healthcare the number of people which will opt for implants will rise continuously, and thereby the demand for continuous innovation in this realm [2,3].

Nowadays, there is an increasing interest focused on coatings for biodegradable Magnesium-based metal implants. The idea of using a biodegradable metallic implant based on Mg is that can rapidly adapt to the body and eventually dissolve after the bone has healed [4,5]. But, besides good properties of Mg, like good biocompatibility, similar mechanical properties to natural bone, or lightweight metallic material, the main disadvantage is his rapidly corrosion rate in physiological environments. Consequently, they degrade faster than the time required for bone healing [6,7]. Consequently, this issue has stimulated numerous research teams around the world to test materials and combinations of materials as coatings on Mg-based metal implants to mitigate the fast corrosion rate [5,8]. The focus of this study concerns the development of biocompatible and biodegradable coatings by pulsed laser deposition (PLD) method of bioactive glasses (BG), on Mg-metallic implants with the intent of reducing and controlling the corrosion rate and increasing their biocompatibility. BG are made of various metal oxides combinations and are superior in bioactivity to calcium phosphates (e.g. hydroxyapatite, octacalcium phosphate) [9–11]. Moreover, after prolonged contact with body fluids *in vivo*, they change their composition transforming in an equivalent of the mineral bone [10,12].

It is widely accepted that laser-based technologies are among the main, most powerful tools for fabrication of micro- and nano-arrays of wide-range of different biomaterials with controlled thickness (with the precision of 1 Å), good adhesion to the substrate and specific surface properties [13]. Up to present time, PLD technique proved to be a versatile technique able to synthesize biomaterials thin films on temporary or biodegradable metallic substrates with improved corrosion properties and excellent biocompatibility [14–18]. The obtained results hint toward the development of alternative viable solutions for a new generation of biodegradable metallic implants with increased functionality.

2. Materials and Methods

2.1 Materials and experiment

Two bioactive glass formulations have been employed as PLD target in this study: 6P57 (wt.%: SiO_2-56.5%, Na_2O-11%, K_2O-3%, CaO-15%, MgO-8.5%, and P_2O_5-6%) and 6P61 (wt.%: SiO_2-61.1%, Na_2O-10.3%, K_2O-2.8%, CaO-12.6%, MgO-7.2%, and P_2O_5-6%). These two compositions are respectively slightly lower and higher than the threshold value of 60 wt.% SiO_2 which corresponds to a significant change in bioactivity and degradability of BG [9,10]. Two grams from each powder were pressed at 5 MPa in a 20 mm diameter mold (model Specac), and the resulting pellets were heat-treated in a furnace (model Carbolite CF1100), at 650°C/6 h, to reach compactness by eliminating air bubbles and water vapours.

PLD depositions were conducted inside a stainless steel deposition chamber using a KrF* excimer laser source, model *COMPexPro 205* (λ = 248 nm, $\tau_{FWHM} \leq 25$ ns), running at a repetition rate of 10 Hz. The laser beam was oriented with an inclination angle of 45° with respect to the target surface. The ablated material was collected either onto Mg disks (thickness=2 mm) or Si (100) wafers and were placed parallel to the targets (on-axis geometry), at a 40 mm separation distance.

During the experiments, the substrates were heated and maintained at a constant temperature of 450°C using a temperature controller (model PID-EXCEL). The depositions have been carried out at a residual pressure of 10^{-3} Pa and in oxygen atmosphere. The experimental conditions were identical for both deposition substrates. Prior to introduction inside the deposition chamber, in order to eliminate the micro-impurities, the substrates were successively cleaned into an ultrasonic bath in acetone, ethanol and deionised water for 20 min and then blown dry with high purity nitrogen. The laser fluence on target was set at ~2.9 J/cm^2. The targets were continuously rotated with 0.3 Hz and translated along two orthogonal axes to avoid the formation of deep craters in order to insure a uniform deposition. This also allowed exposing a "fresh" surface to the action of every laser pulse. For the growth of one film, 20000 consecutive laser pulses were applied.

2.2 Film thickness determination and morphological examination

The coating thickness was monitored and estimated by profilometry using a Stylus Profiler HP-2 from Ambios Technology. A continuous scan at a rate of 10 μms^{-1} and 1 mm working distance were used. For statistics, the measurements were performed on three identical samples. Morphological characterization was carried out by Scanning Electron Microscopy (SEM) using a FEI Inspect S electron microscope, operating at 20 kV acceleration voltage, in high vacuum, under secondary electron mode.

2.3 Compositional and structural investigation

Compositional analysis was performed by Energy Dispersive Spectroscopy (EDS), with a SiLi EDAX Inc. detector, operated at 20 kV.

The long-range order of the simple BG films was investigated by grazing incidence X-ray diffraction (GIXRD) using a Bruker D8 Advance diffractometer, in parallel beam setting, with Cu K_α (λ=1.15406 Å) incident radiation. The incidence angle was set to 2°, and the scattered intensity was scanned within the 2θ range (15–45)°, with a step size of 0.04°, and 25 s per step.

Fourier Transform Infra-Red (FTIR) spectroscopy was employed for analysing the short-range order and nature of functional groups present in the BG films. The measurements were performed with a Perkin Elmer Spectrum BX II spectrometer, in attenuated total reflection mode (ATR) using a Pike-MIRacle diamond head of 1.8 mm diameter. The spectra were collected in the fingerprint wave numbers range 1300–550 cm^{-1}, by recording 128 individual scans at a resolution of 4 cm^{-1}.

3. Results and Discussion

The profilometry measurements in the case 6P57 and 6P61 coatings deposited in oxygen atmosphere indicated average thickness values in the range of ~4 μm. The coatings deposited in vacuum show different thicknesses depending on the type of BG composition: for 6P57≈7 μm and 6P61≈4 μm (figure 1).

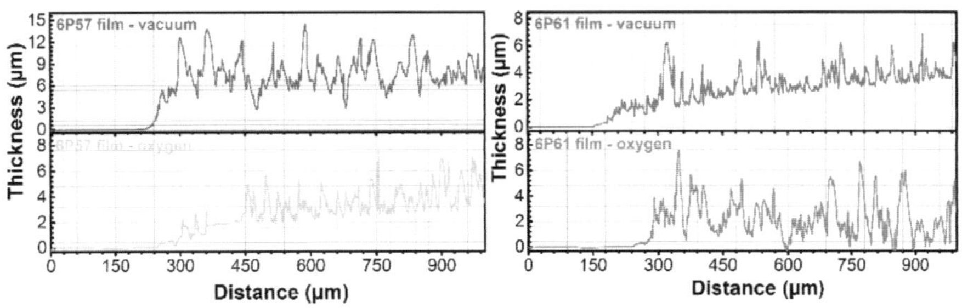

Figure 1. Thicknesses profiles of 6P57 and 6P61 coatings synthesized in oxygen or vacuum atmosphere by PLD method.

The surface morphology of 6P56 and 6P61 coatings deposited in oxygen atmosphere or in vacuum are illustrated in figure 2. It is noted that all coatings show uniform morphology, and consist of a homogenous matrix, consisting of particulates. At higher magnification, these morphological features appeared composed of isolated or merged spheroidal formations, with sizes in the micrometre range and an irregular distribution. Such particulates observed on the top of deposited coatings are characteristic to the PLD process [19]. Many studies demonstrated that these kinds of surfaces are beneficial for cells adhesion and growth [20–22].

Figure 2. SEM images of 6P57 (1a,1b,2c,2d) and 6P61 (a1,b1,c2,d2) coatings deposited onto Si substrates in oxygen (1a,1b,a1,b1) and in vacuum (2c,2d,c2,d2). Panoramic (1a,2c,a1,c2 – magnification bar=300 μm) and detailed (1b,2d,b1,d2 – magnification bar=10 μm) views.

The EDS results for 6P57 and 6P61 coatings are presented in table 1. By comparing BG targets oxides concentration with the obtained coatings, after PLD transfer, it can be observed a good replication of the target materials' composition.

Table 1. Oxide concentration in wt.% of the BG coatings deposited by PLD onto Ti substrates (as estimated based on the quantitative EDS results). Network connectivity (NC) values.

	Oxide concentration (wt. %)						
Sample type	SiO_2	Na_2O	K_2O	CaO	MgO	P_2O_5	NC
6P57-vacuum	56.2±1.3	10.1±0.04	3.2±0.08	21.2±0.4	7.3±0.2	2.0±0.2	2.47
6P57-oxygen	58.2±2.0	10.8±0.06	2.8±0.07	17.1±0.5	7.4±0.2	3.7±0.3	2.73
6P61-vacuum	63.8±4.0	7.6±0.2	2.3±0.1	18.5±1.0	6.8±0.2	1.0±0.2	2.82
6P61-oxygen	61.1±1.8	9.8±0.03	2.5±0.1	15.3±0.5	7.5±0.2	3.8±0.1	2.89

The GIXRD analyses (figure 3) confirmed the amorphous status of the BG coatings regardless of the fabrication conditions.

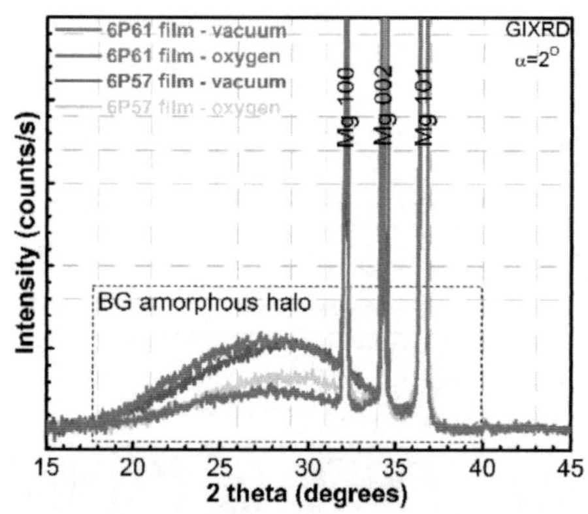

Figure 3. Comparative GIXRD patterns of BG coatings synthesized by PLD in vacuum and oxygen ambient.

The FTIR-ATR spectroscopic investigations evidenced the characteristic molecular vibrations of bonds and functional groups within the structure of either source raw glass powders, PLD targets (sintered at 650°C/6h in air), or BG coatings (figure 4a–c). The silica-based glasses consist of the tridimensional network formed of SiO_4 tetrahedral units (Q_{Si}^n; *n*=0–4, where *n* is the number of bridging oxygen atoms) interconnected at the corners by oxygen bridges. The glass network connectivity decreases with the increase of network modifiers (alkali and alkali-earth elements) concentration, because the charge compensation of the network modifiers occurs at the expense of oxygen bridges breaking. This way it is induced the formation of non-bridging oxygen structural units ($Si - NBO: Q_{Si}^3;\ Q_{Si}^2;\ Q_{Si}^2;\ Q_{Si}^1; and\ Q_{Si}^0$), whom concentration will govern the degradation speed of the glass in aqueous media, such as is the intercellular environment [9,10,23].

All analysed materials (raw powders, PLD targets, and BG coatings) elicited a broad IR band with two prominent shoulders centred at ~942 cm^{-1} (ascribed to the asymmetric stretching vibrations of the Si–NBO units) and at ~1001 cm^{-1} (associated with the asymmetric stretching vibrations of the Si–O–Si bonds). The lower intensity band, situated at ~776 cm^{-1}, is determined by the bending vibrations of the Si–O bonds [24,25]. Due to the fairly low phosphorous concentration of the BG formulations used in this study, the characteristic vibrations of the phosphate functional groups cannot be easily evidenced, as they appear in the same spectral region [26,27] as the more intense absorption bands of the aforementioned silicate groups.

The comparison of the FTIR spectra of the raw powders and PLD targets (figure 4a) revealed that the sintering process applied at 650°C/6h does to induce compositional and structural modifications of the glass material. As expected, the Si–O–Si/Si–NBO ratio is higher in the case of the 6P61 glass, due to its elevated silica content, and consequently lower concentration of network modifiers.

In the case of both types of BG coatings synthesized under vacuum conditions a more reduced connectivity degree (determined by a higher amount of Si–NBO) has been signalled (figure. 4b,c). When oxygen is used as working gas during the PLD fabrication experiments an opposite tendency is noticed: both types of coatings experiencing an increase in intensity of the Si–O–Si band (figure 4b,c).

Thus, the preliminary studies have indicated so far that the control over the BG coatings' degree of network connectivity might be achieved by engineering the PLD working parameters.

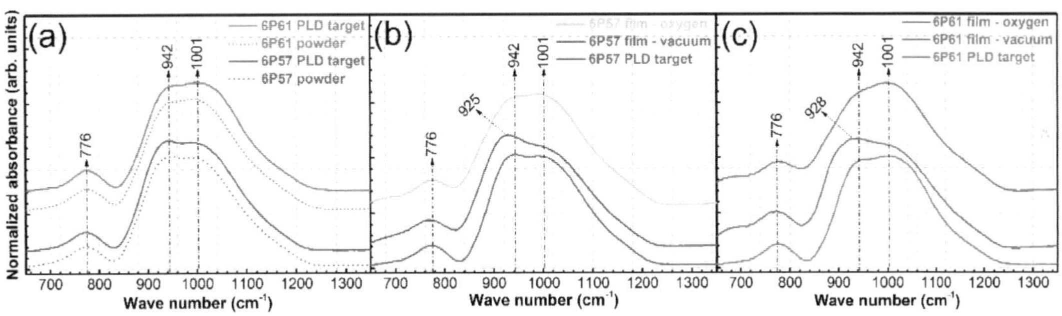

Figure 4. Comparative FTIR spectra of: (a) BG raw source powders and sintered PLD targets; (b) 6P57 PLD target and coatings; and (c) 6P61 PLD target and coatings.

4. Conclusions

BG coatings, with different concentration of silica were deposited on pure Mg and Si substrates, at a temperature of 450°C, by PLD technique in view of developing implants with controlled biodegradability. Preliminary morphological, compositional and structural assessments of the deposited coatings were conducted.

It was shown that the sintering process applied at 650°C/6h to PLD targets does not induced compositional and structural modifications of the glass material. The BG films displayed a rough morphology, suitable for osteointegration applications. The amorphous nature of all synthesized bioactive glass coatings was demonstrated. A better structure conservation was revealed in the case of films deposited under oxygen ambient. EDS has indicated the PLD method capacity to transfer near-congruently the complex composition of BGs on substrates.

We consider that the physico-chemical results recommend the further research of BG coatings fabricated by pulsed laser advanced techniques, as promising biomaterials fit to be integrated in biodegradable metallic implants.

Acknowledgements

N.M., C.R., M.S., and I.N.M. acknowledge with thanks the financial support of UEFISCDI under the contract TE140/2015. G.E.S. thanks the NIMP Core Programme PN 16 48-3/2016.

References

[1] http://kidshealth.org/en/parents/broken-arm.html;

[2] Floroian L, Ristoscu C, Mihailescu N, Negut I, Badea M, Ursutiu D, Chifiriuc M C, Urzica I, Dyia H M, Bleotu C and Mihailescu I N 2016 *Molecules* **21** 740
[3] http://www.marketsandmarkets.com/Market-Reports/orthopedic-device-280.html
[4] Adeosun S O, Lawal G I and Gbenebor O P 2014 *J. Miner. Mater. Charact. Eng.* **2** 88
[5] Bita A I, Stan G E, Niculescu M, Ciuca I, Vasile E and Antoniac I 2016 *J. Adhes. Sci. Technol.* **30** 1968
[6] Li Z, Gu X, Lou S and Zheng Y 2008 *Biomaterials* **29** 1329
[7] Hermawan H 2012 *Biodegradable Metals: From Concept to Applications* (Springer-Verlag Berlin Heidelberg)
[8] Hornberger H, Virtanen S and Boccaccini A R 2012 *Acta Biomater.* **8** 2442
[9] Hench L L 1991 *J. Am. Ceram. Soc.* **74** 1487
[10] Hench L L and Wilson J 1993 *Advanced Series in Bioceramics: Volume 1* (World Scientific Publishing Company, Singapore)
[11] Popa AC, Stan GE, Enculescu M, Tanase C, Tulyaganov DU and Ferreira JMF 2015 *J. Mech. Behav. Biomed. Mater.* **51** 313
[12] Floroian L, Popescu A, Serban N and Mihailescu I 2011 Polymer-bioglass composite coatings: A promising alternative for advanced biomedical implants *Metal, Ceramic and Polymeric Composites for Various Uses* ed. J Cuppoletti (Rijeka InTech) chapter 20 pp 393–420
[13] Pique A 2007 Deposition of polymers and biomaterials using the matrix-assisted pulsed laser evaporation (MAPLE) process *Pulsed Laser Deposition of Thin Films: Applications-Led Growth of Functional Materials* ed. R Eason (John Wiley & Sons, Hoboken NJ) chapter 3 pp 63–82
[14] Nelea V, Jelinek M and Mihailescu IN 2007 Biomaterials: New issues and breakthroughs for biomedical applications *Pulsed Laser Deposition of Thin Films: Applications-Led Growth of Functional Materials* ed. R. Eason (John Wiley & Sons, Hoboken NJ) chapter 18 pp 421–456
[15] Floroian L, Sima F, Florescu M, Badea M, Popescu A C, Serban N and Mihailescu I N 2010 *J. Electroanal. Chem.* **648** 111
[16] Mihailescu I N and Gyorgy E 1999 Pulsed Laser Deposition: An Overview *International Trends in Optics and Photonics* ed. T. Asakura (Springer-Verlag Berlin Heidelberg) chapter 13 pp 201-214
[17] Rau J V, Antoniac I, Fosca M, De Bonis A, Blajan A I, Cotrut C, Graziani V, Curcio M, Cricenti A, Niculescu M, Ortenz M and Teghil R 2016 *Mater. Sci. Eng. C* **64** 362
[18] Floroian L, Savu B, Stanciu G, Popescu A C, Sima F, Mihailescu I N, Mustata R, Sima L E, Petrescu S M, Tanaskovic D and Janackovic D J 2008 *Appl. Surf. Sci.* **255** 3056
[19] Eason R 2007 *Pulsed Laser Deposition of Thin Films: Applications-Led Growth of Functional Materials* (John Wiley & Sons, Hoboken NJ)
[20] Brunette D M, Tengvall P, Textor M and Thomsen P 2001 *Titanium in Medicine: Material Science, Surface Science, Engineering, Biological Responses and Medical Applications* (Springer, Berlin)
[21] Popa M, Hussien M D, Cirstea A, Grigore R, Lazar V, Bezirtzoglou E, Chifiriuc C, Sakizlian M, Stavropoulou E and Bertesteanu S 2015 *Curr. Top. Med. Chem.* **15** 1614
[22] Mihailescu N, Stan G E, Duta L, Chifiriuc M C, Bleotu C, Sopronyi M, Luculescu C, Oktar F N and Mihailescu I N 2016 *Mater. Sci. Eng. C* **59** 863
[23] O'Donnell M D 2012 Melt-Derived Bioactive Glass *Bio-Glasses: An Introduction* ed. JR Jones and A G Clare (John Wiley & Sons, Ltd, Chichester, UK) chapter 2 pp 13–27
[24] Popa A C, Marques V M F, Stan G E, Husanu M A, Galca A C, Ghica C, Tulyaganov D U, Lemos A F and Ferreira J M F 2014 *Thin Solid Films* **553** 166
[25] Popa A C, Stan G E, Besleaga C, Ion L, Maraloiu V A, Tulyaganov D U and Ferreira J M F 2016 *ACS Appl. Mater. Interfaces* **8** 4357

[26] Stuart B, Gimeno-Fabra M, Segal J, Ahmed I and Grant D M 2015 *ACS Appl. Mater. Interfaces* **7** 27362
[27] Stuart B W, Titman J J, Gimeno-Fabra M, Ahmed I and Grant D M 2016 *Mater. Des.* **111** 174

Oxide/ metal/oxide nanolaminate structures for application of transparent electrodes

Hr Dikov, T Ivanova and P Vitanov
Central Laboratory of Solar Energy and New Energy Sources,
Bulgarian Academy of Sciences, Blvd. Tzarigradsko chaussee 7, Sofia 1784, Bulgaria

E-mail: vitanov@phys.bas.bg

Abstract. Transparent and conductive oxide/ metal/ oxide nanolaminate structures were deposited on glass and polymer substrate by RF magnetron sputtering without substrate heating. The Ag nanoparticles with different size and distance between neighboring particles were located on the interface of two thin oxide layers. This sputtering configuration allows obtaining thin films with homogeneous thickness. The three targets give the opportunity to deposit successively three different layers without opening the chamber. The developed process for transparent conducting coating is a low temperature and it is suitable for application on organic materials as substrate and foils. The experiment with different substrates manifest that the optical transparency of the conducting coating depends on substrate material. The obtained results have demonstrated that the nanolaminate structures oxide/metal/oxide (OMO) as TCO coating are especially suitable for applications in flexible electronics and optoelectronics

1. Introduction

In general, TCOs are degenerated n-type semiconductors with intrinsic doping by native donors such as oxygen vacancies and/or interstitial metal atoms and additional extrinsic doping by donor impurities. There is an inherent limitation in the metal oxide conductivity that can be obtained by increasing the carrier concentration, because the Coulomb interaction between the free electrons and the ionized donor centers from which they are generated provides a source of scattering that is inherent to the doped material. In addition, for metal oxides with a high number of free carriers, some absorption of the incident radiation by interaction with the electron gas takes place around the characteristic electron plasma frequency which increases with increasing carrier concentration. When the densities become greater than 2×10^{21} cm^{-3}, the TCO exhibits plasma frequencies that shift from absorbing infrared wavelengths to visible light, reducing the transparency in the visible region [1].The requirement of transparency and the fundamental scattering mechanism establish an absolute limit to the TCO resistivity of about $4\times10^{-5}\Omega$.cm or obtaining conductivity in such materials, they should be doped with an appropriate element. Recently, it was reported for the preparation of an optically transparent and conductive nanolaminated dielectric structure [2].The idea is based on usage of electronic conductivity in granular (discontinuous) type materials [3]. The granules are metallic particles of sizes ranging usually from a few to hundreds of nanometers, embedded into an insulating matrix. The nanolaminate structure is formed using one or two different dielectric materials.

In present study, transparent and conductive oxide/ metal/ oxide nanolaminate structures were deposited on glass and polymer substrate by RF magnetron sputtering without intentional substrate heating. The Ag nanoparticles were located on the interface of two thin oxide layers. The observed optical and electrical properties were dependent on the substrates. The nanolaminate structures showed high optical transmittance and good electrical conductivity.

2. Experimental

RF magnetron sputtering system with three different targets has been used for the deposition of the layers. The sputter gas is argon. This system has been developed as low- temperature high-speed sputtering equipment for use in experiments and is suitable for thin film deposition of metals and insulating materials. The substrate holder has a size of 200 mm in diameter. The three targets (75 mm in diameter) are at the eccentric position. They are cooled with water (3 l /min), so little radiation heat is generated. The targets and the substrate holder are vertically arranged at a distance of 8 cm. The substrate holder is spinning with a rate of 10 - 80 rpm. This configuration allows obtaining thin films with a homogeneous thickness. The three targets give the opportunity to deposit successively three different layers without opening the chamber. The most important part of the deposition process of nanolaminate structure with specified optical and electrical properties is the deposition of discontinuous metal layer. The sizes and density of the metal granules depend on the duration of the sputtering, RF power, type of reactive gas (Ar or He) and the rotation rate of substrate holder.

The optical properties of the nanolaminated structure dielectric/metal/dielectric (oxide/metal/oxide-OMO) are studied by means of UV-VIS-NIR Shimadzu spectrophotometer UV 3300. The sheet resistance has been measured using the four- point probe method.

3. Results and discussions

The studied nanolaminate structures are Oxide/metal (Ag)/Oxide layers (OMO). The investigations reveal that the relation between the sheet resistance and the maximum transparency in the visible spectral range is depending on the size of the metal granules, the film thickness of the dielectric layer, the type of the oxide, which is used for the formation of the OMO structure.

3.1. Influence of the thickness of the oxide layers

The nanolaminate structures TiO_2/metal (Ag)/TiO_2 have been investigated by keeping the 10 s sputtering time of Ag (10 s), but varying the sputtering time of TiO_2 layers. The glass substrate is 1.4 mm thick. Table 1 represents the sheet resistance values for nanolaminate structures with different thickness of oxide layer.

Table 1. Sheet resistance of nanolaminate structures varying TiO_2 thickness

Structure	Sheet resistance [Ω/□]
Sample 1- TiO_2 (1 min)/10s Ag/TiO_2 (3 min)	12
Sample 2- TiO_2 (1 min)/10s Ag/TiO_2 (2 min)	11
Sample 3 - TiO_2 (3 min)/10s Ag/TiO_2 (3 min)	10

The sheet resistance has not considerable changed with varying the thickness of the consisting metal oxides. On the other hand, the transmittance is changed with changing the film thickness of TiO_2 layer. Interestingly, the best transparency is manifested by the nanolaminate structure with thickest bottom and top TiO_2 layer and the transmittance is 90 % at wavelength 550 nm, the sample 1 and 2 possess values of 78 and 82 % at 550 nm, respectively. The highest reflectance is observed for sample 2- TiO_2 (1 min)/10s Ag/TiO_2 (2 min) (see figure 1).

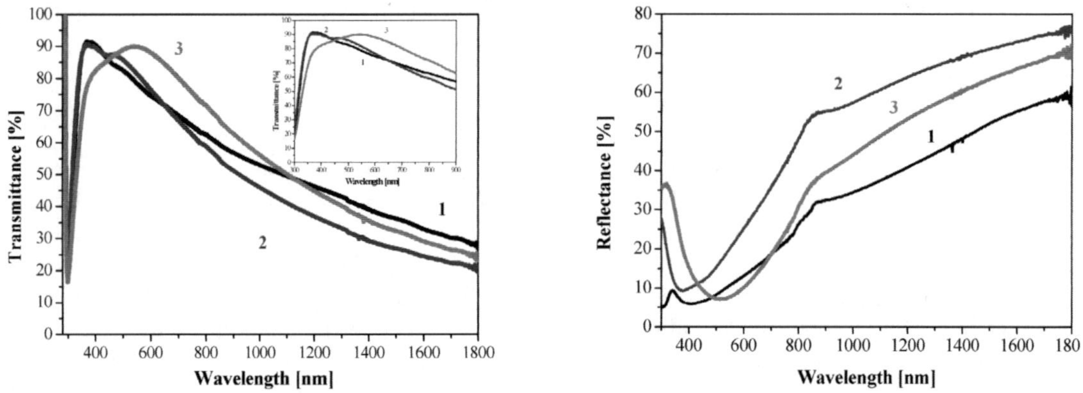

Figure 1. Transmittance and reflectance spectra of samples 1, 2 and 3, described in table 1. Transmittance is measured using bare glass substrate as background..

The effect of the layer of Ag nanoparticles is found to be similar for other nanolaminate structure, using different metal oxide. The studied structures are MoO_x/ Ag /MoO_x, where molybdenum oxide films are deposited by magnetron sputtering, using MoO_2 target in Ar atmosphere. The RF power is 150 W for deposition of MoO_x and 200 W for deposition of Ag nanoparticles. There are studied two structures and pure MoO_x layer.

Table 2. Sheet resistance of nanolaminate structures varying MoO_3 thickness

Structure	Sheet resistance [Ω/□]
Sample 1- MoO_x (90 s)/10s Ag/MoO_x (90s)	7
Sample 2- MoO_x (90 s)	280
Sample 3 –MoO_x(45 s)/10s Ag/MoO_x (45s)	10

* The labels 90s and 45s mean the deposition time

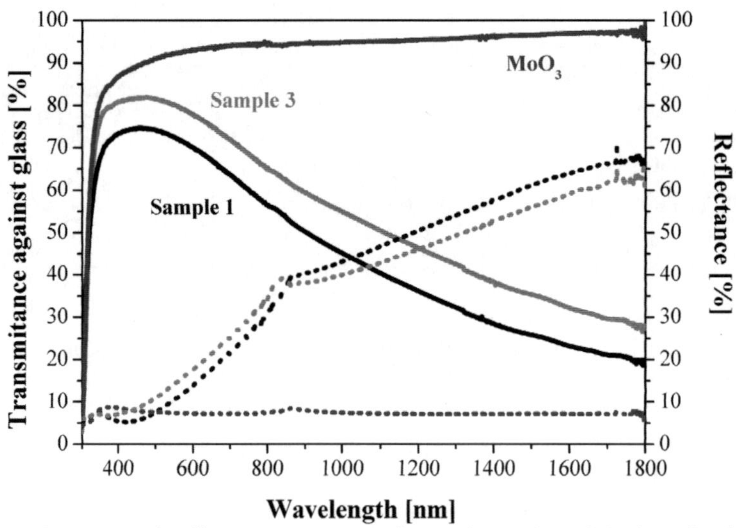

Figure 2. Transmittance and reflectance spectra of samples 1, 2 and 3, described in table 2 of MoO_x and MoO_x (90s)/ Ag (10s)/MoO_3 (90s). Transmittance is measured using bare glass substrate as background.

The nanolaminate structure of sample 3 has the same film thickness of MoO_x as the sample 2, but the difference is the presence of intermediate layer with Ag nanoparticles. The pure MoO_x film is very transparent in the visible spectral range above 90 %. As it can be seen from figure 2, the reflection of the multilayer system increased with Ag layer. The increase in the reflection in the near infrared region is due to the interaction of free electrons with the incident radiation.

3.2. Influence of the substrate used

Studies on structures of the type ITO/ Ag / ITO on organic substrates were reported in several papers [4 -6]. The results are related to transparent conducting oxide (TCO) materials containing indium as dopant.

3.2.1. Optical and electrical properties of the nanolaminate structures on Plexiglass

Poly(methyl methacrylate) (PMMA), also known as acrylic glass as well as by the trade name Plexiglas, is a transparent thermoplastic often used in sheet form as a lightweight or shatter-resistant alternative to glass substrate. The used PMMA substrate is 1.5 mm thick, the deposited nanolaminate structure is TiO_2 (6 nm)/Ag NPs/TiO_2 (12 nm). The measured sheet resistance of this structure is 9 Ω/□.

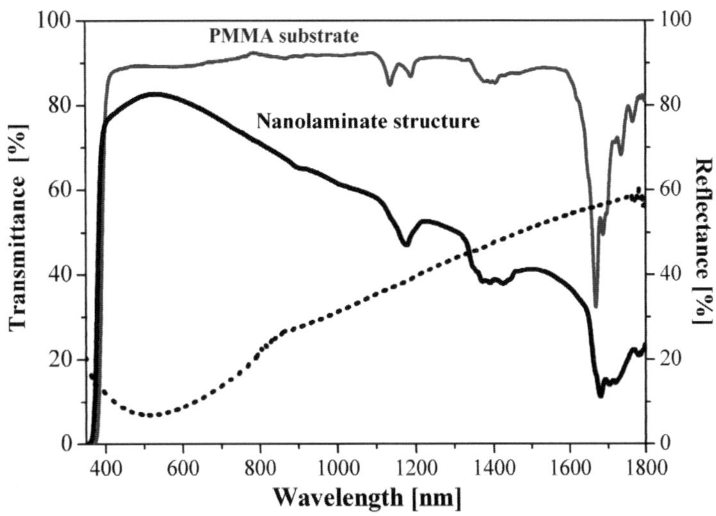

Figure 3. Transmittance and reflectance spectra of TiO_2 (6 nm)/Ag NPs/TiO_2 (12 nm). The transmittance is measured against air (solid lines) and the reflectance is dotted line, The spectrum of bare PMMA substrate is given only for comparison.

The transmittance of the nanolaminate structure is near 80% in the visible spectral range and drops significantly for wavelength above 800 nm, meanwhile the reflectance below 10 % up to 700 nm and then increased in the near IR spectral region.

3.2.2. Optical and electrical properties of the TiO_2 / Ag / TiO_2 structure on flexible substrates

Nanolaminate structure of type TiO_2 (20 nm)/ Ag (NPs) /TiO_2 (20 nm) is sputtered simultaneously on four types substrates including glass substrate, two substrates of PVC foil with thickness 0.3 µm, and 0.15 µm and Hostaphan foil with thickness 0.1 µm. Hostaphan® is a family of PET (polyester) foils. The measured sheet resistance is presented in table 3.

The structures on PVC substrates have a greater sheet resistance in comparison with the glass substrate due to the rough surface of the material. Perhaps in this case the uniform distribution of the silver nano-granules is disrupted by increasing the distance between them. Structures deposited on

PVC substrates have higher sheet resistance in comparison with the same structures on glass substrates,(see Table 3). This may be due to rough surface of the PVC material. Probably, in this case the uniform distribution of the silver nano-granules is disrupted with increasing the distances among Ag granules.

Table 3. Sheet resistance of nanolaminate structure TiO_2 /(NP) Ag/TiO_2 on different types of substrates

Structure	Sheet resistance [Ω/□]
Sample 1- on glass substrate	8
Sample 2- structure on PVC foil 0.3 μm	285
Sample 3 - PVC foil, 0.15 μm	375
Sample 4 - Hostaphan foil, 0.1 μm	8

Figure 4 shows that the layers on the PVC substrate have smaller transparency compared to those on glass. The transmittance spectra is measured using the relevant substrate as background. Optical and electrical properties of the nanolaminate structure on Hostaphan foil do not differ substantially from the sample on glass.

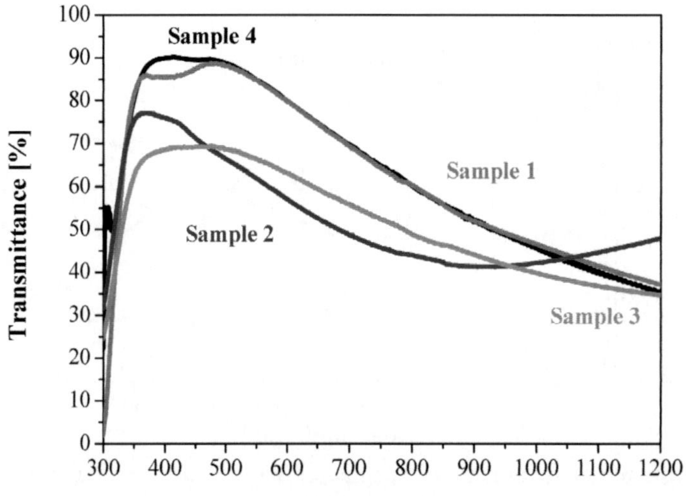

Figure 4. Transmittance spectra of TiO_2/ Ag NPs/TiO_2 on four types of substrates.

Table 4. Sheet resistance of nanolaminate structures on Hostaphan foil 0.1 μm thick substrate.

Structure	Sheet resistance [Ω/□]
Sample 1-TiO_2(20nm)/Ag(NPs)/ TiO_2(15nm)	7.5
Sample 2- TiO_2 (10 nm)/Ag(NPs)/ TiO_2 (20nm)	8.0
Sample 3 - TiO_2 (20nm)/Ag(NPs)/ TiO_2 (20 nm)	8.0

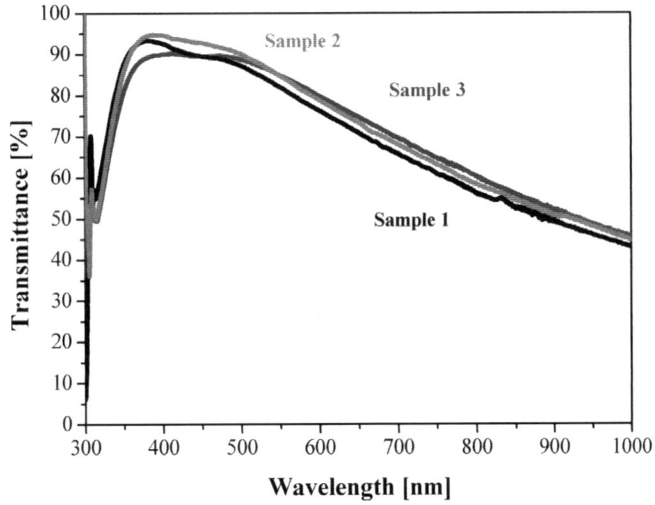

Figure 5. Transmittance spectra of nanolaminate structures decribed in table 4 deposited on Hostaphan foil 0.1μm thick substrate.

To demonstrate the possibilities for optimization of the properties of nanolaminatnata structure as the TCO layer, samples with different thicknesses of TiO_2 were tested. The electrical conductivity did not change significantly (see Table 4), but the effect on the transmittance spectra was observed. Figure 5 shows the transmittance spectra of three different nanolaminate structures deposited on Hostaphan foil with thickness 0.1 μm. It can be seen that they possess almost the same transparency.

4. Conclusions

The developed method for transparent conducting coating is a low temperature process and it is suitable for application on organic materials as substrate and foils. The experiment with different substrates manifest that the optical transparency of the conducting coating depends on substrate material. One suitable material is proved to be Hostaphan. The obtained results have demonstrated that the nanolaminate structures oxide/metal/oxide (OMO) as TCO coating are especially suitable for applications in flexible electronics and optoelectronics.

5. References

[1] Knickerbocker S A and Kulkani A K 1996 *J. Vac. Sci. Technol. A* **14** 757
[2] Dikov Hr, Vitanov P, Ivanova T and V Stavrov 2016 *IOP: Conference Series* **700** 012054
[3] Beloborodov I, Lopatin A, Vinokur V and Efetov K 2007 *Rev. Mod. Phys.* **79** 469
[4] Park J S, Choi K H and Kim H K 2009 *J. Phys. D: Appl. Phys.* **42** 235109
[5] Choi K H, Nam H J, Jeon J A, Cho S W, Kim H K, Kang J W, Kim D C and Cho W J 2008 *Appl. Phys. Lett.* **92** 223302
[6] Guillen C and Herrero J 2011 *Thin Solid Films*

Magnetron Sputtered Molybdenum Oxide for Application in Polymers Solar Cells

M Sendova-Vassileva [1*], Hr Dikov [1], P Vitanov [1], G Popkirov [1], R Gergova [1], G Grancharov [2] and V Gancheva [2]

[1] Central Laboratory of Solar Energy and New Energy Sources, Bulgarian Academy of Sciences, 72 Tzarigradsko Chaussee, 1784 Sofia, Bulgaria
[2] Laboratory of Structure and Properties of Polymers, Institute of Polymers, Bulgarian Academy of Sciences, Acad. G. Bonchev St., Block 103-A, 1113 Sofia, Bulgaria

*E-mail: marushka@phys.bas.bg

Abstract. Thin films of molybdenum oxide were deposited by radio frequency (RF) magnetron sputtering in Ar from a MoO_3 target at different deposition power on glass and silicon substrates. The thickness of the films was determined by profilometer measurements and by ellipsometry. The films were annealed in air at temperatures between 200 and 400°C in air. The optical transmission and reflection spectra were measured. The conductivity of the as deposited and annealed films was determined. The crystal structure was probed by Raman spectroscopy. The oxidation state of the surface was studied by X-ray photoelectron spectroscopy (XPS) spectroscopy. The deposition technique described above was used to experiment with MoOx as a hole transport layer (HTL) in polymer solar cells with bulk hetrojunction active layer, deposited by spin coating. The performance of these layers was compared with poly(3,4-ethylenedioxythiophene):polystyrene sulfonate (PEDOT:PSS), which is the standard material used in this role. The measured current-voltage characteristics of solar cells with the structure glass/ITO/HTL/Poly(3-hexyl)thiophene (P3HT):[6,6]-phenyl-C61-butyric acid methyl ester (PCBM)/Al demonstrate that the studied MoOx layer is a good HTL and leads to comparable characteristics to those with PEDOT:PSS. On the other hand the deposition by magnetron sputtering guarantees reliable and repeatable HTLs.

1. Introduction

Organic polymer solar cells are the subject of intense research activity with the aim of increasing their efficiency and stability. The most widely used kind of polymer solar cells is that based on the so called bulk heterojunction where the donor and the acceptor are intermixed in a single layer. The functioning of such devices depends on the use of selective layers – hole transport (HTL) and electron transport layers, which facilitate the extraction of one type of charge carrier and present a barrier to the other. The typical HTL is PEDOT:PSS. Because of some problems that the use of PEDOT:PSS presents connected with the stability of the cells there has been a search for alternatives. A number of inorganic oxides have been experimented and one of them is molybdenum oxide. It has been successfully applied in both normal and inverted polymer solar cells [1,2] on glass as well as on flexible substrates [3,4]. Molybdenum oxide is found to have a high work function, which is close in value to the highest occupied molecular orbital (HOMO) of a lot of the donor materials in bulk heterojunction polymer solar cells like P3HT, Poly[N-9'-heptadecanyl-2,7-carbazole-alt-5,5-(4',7'-di-2-thienyl-2',1',3'-

Journal of Physics: Conference Series **764** (2016) 012022 doi:10.1088/1742-6596/764/1/012022

benzothiadiazole)] (PCDTBT), etc. [5,6]. This fact makes it a good hole transport material while the barrier for electrons in the lowest unoccupied molecular orbital (LUMO) of the acceptor is high. The oxide layer is quite thin, on the order of 5 -10 nm.

The most often used method for the deposition of molybdenum oxide as HTL is thermal evaporation [7]. There are a few studies of sputtered films for this purpose [8,9]. In all cases however the sputtering is done from a metal target in reactive O_2+Ar atmosphere. In this contribution molybdenum oxide layers obtained by magnetron sputtering of a MoO_3 target in pure Ar are presented and it is shown that they can be successfully employed as hole transport layers.

2. Experimental

Thin films were deposited by magnetron sputtering of a MoO_3 target (76.2 mm dia. x 3.18 mm thick, purity 99.99%) purchased from Testbourne Ltd. It was water cooled during the deposition. The sputtering was performed in Ar atmosphere for 10 min and the RF power was varied between 100 W and 250 W. The substrates were glass (76x26 mm) and 2" Si wafers. The target and substrate holder are situated vertically at a distance of 8 cm from each other. The substrate holder is rotated at a rate of 80 rpm/min during the deposition which leads to greater homogeneity of the films. The deposition rate was between 35 and 90 Å/min depending on the RF power. Some of the films on glass were annealed for 1 hour in air at 200 °C, 300 °C and 400 °C.

The thickness of the films was measured by a Talystep profilometer. The optical transmission and reflection of the films was measured with a Shimadzu UV-3600 spectrophotometer. The Raman spectra were measured on glass substrates with a Horiba Jobin Yvon LabRAM HR800 spectrometer using the 600 l/mm grating and HeNe laser for excitation. The samples were placed under the 100X achromatic objective of a BX41 microscope and measured in back scattering configuration. The sheet resistance was measured using the four point method. The XPS spectra were recorded with an AXIS Supra spectrometer of Kratos Analytical Ltd. and fitted using CasaXPS software package. The background intensity for scans was calculated using a Shirley background substitution.

In order to test the applicability of the sputtered MoOx films as HTL in polymer bulk heterojunction (BHJ) solar cells the following devices were prepared. On structured ITO covered glass substrates from Ossila Ltd. was deposited either PEDOT:PSS by spin-coating or MoOx by RF magnetron sputtering at 150 W for 3 min. to get a 15 nm thick layer. On top the P3HT:PCBM BHJ was spin-coated. Al back contacts were evaporated. The devices were annealed after encapsulation at 110 °C for 10 min. on a hot plate. Their current-voltage characteristics were measured using a halogen lamp and computer controlled equipment. The illumination used was not standard so the measured characteristics cannot be directly compared to those in the literature. However all solar cells were measured under identical illumination conditions and can be compared between themselves.

3. Results and Discussion

3.1. Optical Properties

The optical transmission of the as deposited films is shown in figure 1. All films have a greyish colour. The difference in the transmission between the films sputtered at 100 and 150 W is connected to the difference in thickness. The film sputtered at 250 W is darker.

Figure 2 presents the Tauc plots for assessing the optical band gap of the as deposited films. It is not possible to determine very accurately the optical gap of these disordered not fully stoichiometric films but the values are within the range cited in the literature for MoOx layers [7,10].

Annealing in air improves the transparency of the layers as demonstrated by the data in figure 3 for a film deposited at 150W. The best results are achieved after annealing at 300 °C. This annealing temperature is too high if the layers are to be deposited on top of the polymer active layer but could be used in the case when the HTL layer is deposited before the organic one.

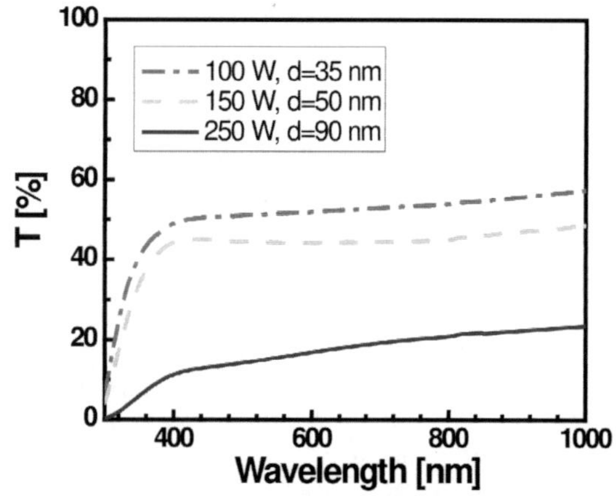

Figure 1. Optical transmission of MoOx films deposited by magnetron sputtering at different power. The thickness of the films is indicated in the figure.

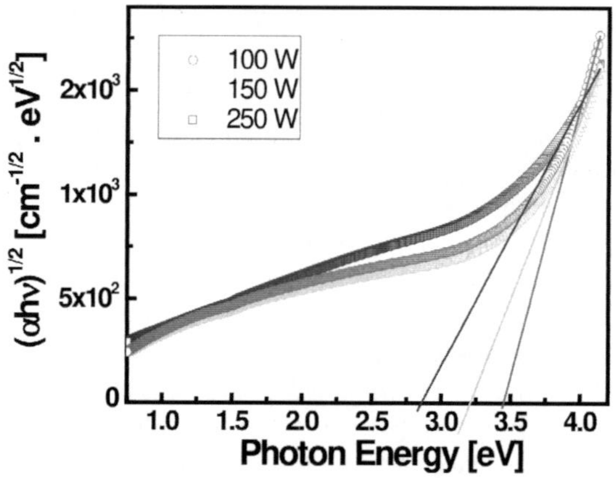

Figure 2. Optical band gap determination of the same films as in the previous figure.

3.2. Raman Spectra

Raman spectra (figure 4) show that the as deposited films are amorphous. Annealing above 300 ºC leads to the appearance in the Raman spectra of a number of sharp lines. After annealing at 400 ºC in most cases the majority of the lines in the Raman spectrum can be assigned to the orthorhombic α-MoO_3 crystalline phase [11,12] (figure 4).

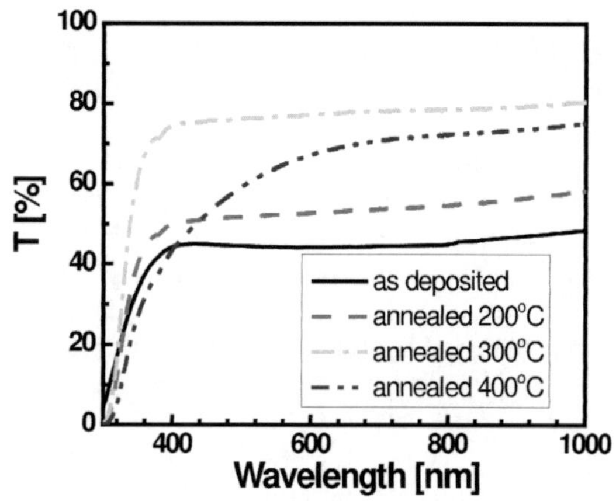

Figure 3. Optical transmission of the film deposited at 150 W as a function of annealing temperature as indicated in the figure.

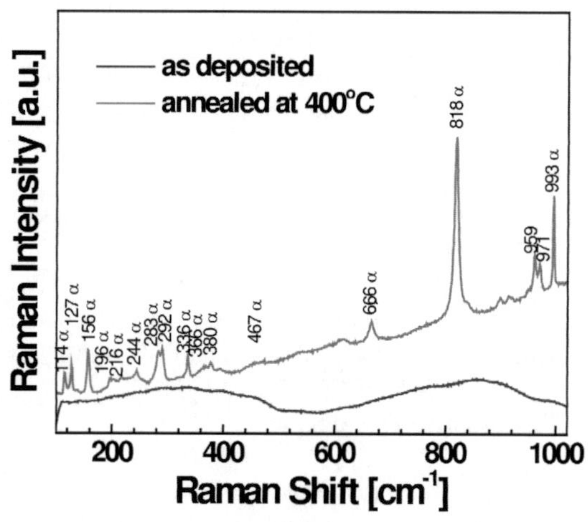

Figure 4. Raman spectra of an as deposited MoOx layer and one annealed at 400°C (upper curve). The lines marked by α are assigned to orthorhombic α-MoO_3.

3.3. Conductivity Measurements

The measured sheet resistance of the as deposited and annealed films is shown in table 1. The conductivity has a metal character, the type of majority carrier could not be determined.

Table 1. Values of the sheet resistance of MoOx films before and after annealing (Ω/□).

Annealing	Deposition power (W)		
	100 W	150 W	250 W
as deposited	540 Ω/□	680 Ω/□	680 Ω/□
200°C	540 Ω/□	680 Ω/□	680 Ω/□
300°C	250 Ω/□	-	260 Ω/□
400°C	260 Ω/□	260 Ω/□	260 Ω/□

3.4. XPS Spectra

The XPS spectra of the surface of the films deposited on silicon were measured. They do not vary much with the deposition power and those of the sample deposited at 150W are shown in figure 5 for Mo3d electrons and in figure 6 for O1s electrons. The information extracted from the spectra after fitting is given in table 2 and table 3 respectively. The content of the different oxidation states of molybdenum is derived from the relative area of the corresponding XPS Mo3d doublet peaks obtained from the fitting. A similar procedure is used for the bonds of oxygen in the O1s XPS spectrum. The fits using the CasaXPS software are very accurate and the content can be evaluated to a precision better than 0.1 at. %. On the basis of the content of the different oxidation states of molybdenum the O:Mo ratio in the molybdemun oxide is calculated presuming that all molybdenum ions in the sampled region are accounted for by the Mo3d XPS spectra.

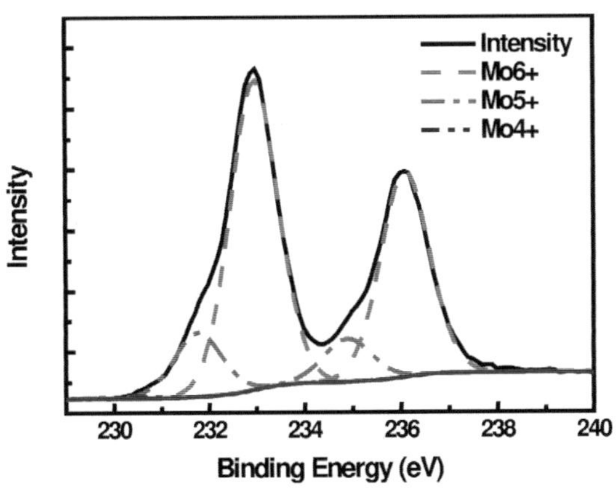

Figure 5. Mo3d XPS spectrum of the sample sputtered at 150W.

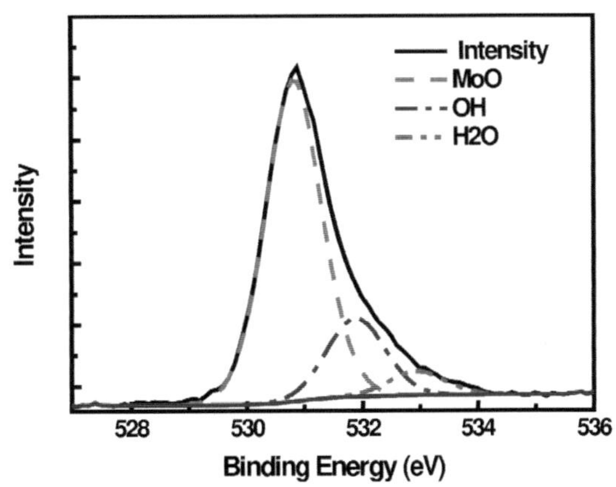

Figure 6. O1s XPS spectrum of the sample sputtered at 150W.

Table 2. Data derived from the Mo3d XPS spectra of the 150W sample.

Mo3d			
	Mo^{6+}	Mo^{5+}	Mo^{4+}
$3d^{5/2}$ position (eV)	236.1	234.9	233.6
$3d^{3/2}$ position (eV)	233.0	231.8	230.4
content	82.5%	17.1%	0.4%
O:Mo	2.91		

The presented XPS data show that the as deposited material is just a little bit substoichiometric molybdenum oxide MoOx with x=2.91 and predominantly 6+ and some 5+ oxidation states of Mo, traces of 4+ and no lower oxidation states. There is some water and –OH groups on the surface which is normal for layers kept in ambient atmosphere. These values for the oxidation state of molybdenum in the as deposited amorphous films correspond to data in the literature for good HTL molybdenum oxide layers [7].

Table 3. Data derived from the O1s XPS spectra of the 150W sample.

O1s			
	Mo-O	**OH^-**	**H_2O**
position (eV)	530.8	531.9	532.9
content	75.8%	18.5%	5.7%

3.5. Results for Application in Solar Cells

The measured current voltage characteristics under identical illumination of the two types of solar cells with the structure glass/ITO/HTL/P3HT:PCBM/Al, differing only in the HTL layer – either MoOx sputtered at 150 W, or PEDOT:PSS deposited by spin-coating, are presented in figure 7. The MoOx layer has not been optimized for this device in any way. In spite of this, the results for the solar cell with the oxide HTL are comparable to those with PEDOT:PSS. This is a positive result showing that the films sputtered from a MoO_3 target only in argon can be employed instead of evaporated or reactively sputtered films but further optimization of the thickness and of the deposition and treatment conditions are needed.

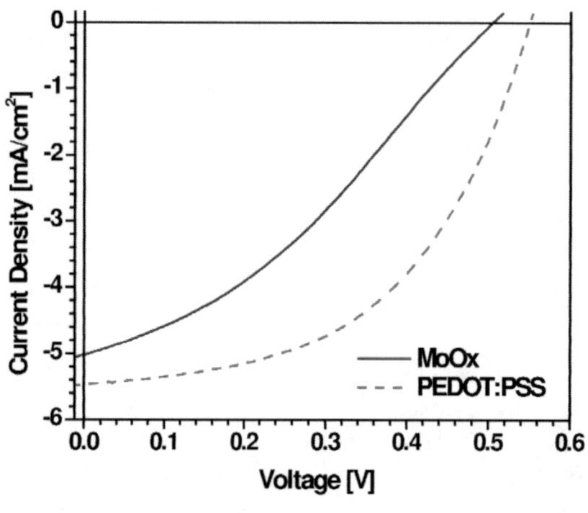

Figure 7. Current-voltage characteristics of solar cells with two different hole transport layers – PEDOT:PSS and MoOx.

4. Conclusion

The results from this study demonstrate that samples deposited by RF magnetron sputtering in pure Ar from a MoO_3 are amorphous. Their transparency in the visible is not very good but good enough for films of about 10 nm thickness to be used as HTL in polymer solar cells. The optical gap estimated from the optical transmission and reflection spectra is around 3 eV and corresponds to values found for amorphous molybdenum oxide in previous studies. According to XPS measurements the as deposited samples are suboxide MoOx with x=2.91 Oxidation states lower than 5+ are found only as traces on the surface. The samples exhibit high planar conductivity.

Polymer solar cells with not optimized sputtered MoOx hole transport layer demonstrate comparable J-V parameters to those of similar cells with PEDOT:PSS HTL. Further optimization of the oxide hole transport layer is needed. However this study demonstrates that magnetron sputtering in pure Ar from an MoO_3 target can be used instead of reactive sputtering or thermal evaporation to obtain MoOx layers for application in solar cells.

Acknowledgement
The authors are grateful to the Bulgarian Fund for Scientific Research (Contract DFNI E02/5 of 12.12.2014) for research funding and to Project INERA (REGPOT-2012-2013-1 NMP, EU 7FP) for financing of conference participation.

References

[1] Bovill E S R, Griffin J, Wang T, Kingsley J W, Yi H, Iraqi A, Buckley A R and Lidzey D G 2013 *Appl. Phys. Lett.* **102** 183303
[2] Cha H-C, Huang Y-C, Hsu F-H, Chuang C-M, Lu D-H, Chou C-W, Chen C-Y and Tsao C-S 2014 *Sol. Energ. Mat. Sol. C.* **130** 191–8
[3] Zhao D W, Tan S T, Ke L, Liu P, Kyaw A K K, Sun X W, Lo G Q and Kwong D L 2010 *Sol. Energ. Mat. Sol. C.* **94** 985–91
[4] Wang Z, Zhang C, Chen D, Tang S, Zhang J, Wang Y, Han G, Xu S and Hao Y 2015 *IEEE Photonics J.* **7** 8400109
[5] Matsushima T, Jin G-H, Kanai Y, Yokota T, Kitada S, Kishi T and Murata H 2011 *Org. Electron.* **12** 520–8
[6] Guo Y and Robertson J 2014 *Appl. Phys. Lett.* **105** 222110
[7] Sian T S and Reddy G B 2004 *Sol. Energ. Mat. Sol. C.***82** 375–86
[8] Rouhani M, Foo Y L, Hobley J, Pan J, Subramanian G S, Yu X, Rusydi A and Gorelik S 2013 *Appl. Surf. Sci.* **273** 150–8
[9] Fan X, Fang G, Qin P, Sun N, Liu N, Zheng Q, Cheng F, Yuan L and Zhao X 2011 *J. Phys. D: Appl. Phys.* **44** 045101
[10] Werner J, Geissbühler J, Dabirian A, Nicolay S, Morales-Masis M, De Wolf S, Niesen B and Ballif C 2016 *ACS Appl. Mater. Interfaces* **8** 17260–7
[11] Seguin L, Figlarz M, Cavagnat R and Lassègues J-C 1995 *Spectrochim. Acta A* **51** 1323–44
[12] Klinbumrung A, Thongtem T and Thongtem S 2012 *J. Nanomater.* **2012** 930763

Strong surface enhanced Raman scattering from gold nanoarrays obtained by direct laser writing

V G Ivanov[1*], N D Todorov[2], L S Petrov[3], T Ritacco[2,3], M Giocondo[2], E S Vlakhov[4]

[1] Sofia University, Faculty of Physics, 5 James Bourchier Blvd., 1164 Sofia, Bulgaria
[2] Consiglio Nazionale delle Ricerche - Istituto di Nanotecnologia - Nanotec SS di Cosenza, Ponte P. Bucci Cubo 31C, 87036 Rende (CS), Italy
[3] Università della Calabria – Dipartimento di Fisica, Ponte Pietro Bucci Cubo 33/B, 87036 Rende (CS), Italy
[4] Institute of Solid State Physics, Bulgarian Academy of Sciences, 72 Tzarigradsko Chaussee, Sofia 1784, Bulgaria

E-mail: vgi@phys.uni-sofia.bg

Abstract. We report for surface enhanced Raman scattering (SERS) from arrays of gold nanoparticles produced by 2-photons photo-reduction of the metallic precursor ($HAuCl_4$) hosted in a Poly-Vinyl Alcohol (PVA) matrix, on glass substrates. Samples with the same pattern but featuring different nanoparticles size and density were obtained by varying the writing laser power and scanning speed. The Raman spectra were recorded from samples immersed in a solution of rhodamine-6G (R6G), as well as, after exposure of the samples in xylene. SERS enhancement factors of up to $\sim 10^4$ were obtained for both analytes. The measurements show that the SERS enhancement is maximized on golden strips produced at higher writing laser power and lower scanning speed, where closer nanoparticles packing is obtained..

1. Introduction

Surface-Enhanced Raman Scattering (SERS) is a powerful analytical technique for detection of trace amounts of chemical and biological active molecules adsorbed on rough metallic surfaces or on metallic nanoparticles. Currently, different mechanisms of SERS have been identified [1], the major one being the electrodynamic enhancement of the local electric field due to surface plasmon resonance (SPR) [2,3]. The most favourable conditions for SPR enhancement are met on specific "hot spots" of the surface, which are usually associated with nanoscale metallic tips [4,5] (tip hot spots) or the gap between two closely spaced metallic nanoparticles, the so called "gap hot spots" [6,7]. Much of the efforts in SERS research in recent years have been focused on fabrication of artificial plasmonic structures in order to maximize the electrodynamic enhancement, as well as to accurately tune the SPR in a desired spectral range [8].

The 2-Photon Direct Laser Writing (2-PDLW) is a promising technique for tailoring of 2D and 3D structures on a sub-wavelength scale in photosensitive materials [9]. The method has been successfully applied for fabrication of mechanical and photonic devices [10–12]. Recently 2-PDLW has been utilized for production of 1D periodic gratings of gold nanoparticles (GNP), by the Au^{+3} photo-reduction, in a Poly-Vinyl Alcohol (PVA) matrix doped with $HAuC_4$ [13]. It has been demonstrated that the width of the gold strips can be finely tuned by choosing appropriate writing laser power and scanning speed. More interestingly for SERS applications, however, the GNP diameter distribution and surface density can also be controlled by the laser exposure parameters. Thus, at higher laser power and lower scanning speed larger density of GNPs could be achieved, which eventually can give rise to a formation of gap hot spots due to maximization of the close particle-particle contacts. Here we report for a strong SERS effect from the gold nanoarrays obtained by 2-PDLW on PVA+ $HAuCl_4$ matrix. The SERS properties of the structures have been tested against rhodamine-6G and xylene, and enhancement factors (EF) superior to 10^4 have been quantified.

2. Samples and methods

A batch of 50 gold 1D arrays was produced by 2-PDLW on a glass substrate by following the procedure described in reference [13]. Arrays were organized in a 5×10 matrix of a size 400×600 μm, as schematically shown in figure 1, with the laser power increasing in the range 2.5-22.5 mW along the columns with a step of 5 mW, and the scanning speed increasing from 10 to 100 μm/s along the rows with a step of 10 μm/s. The period of the parallel gold strips within a single array is 4 μm.

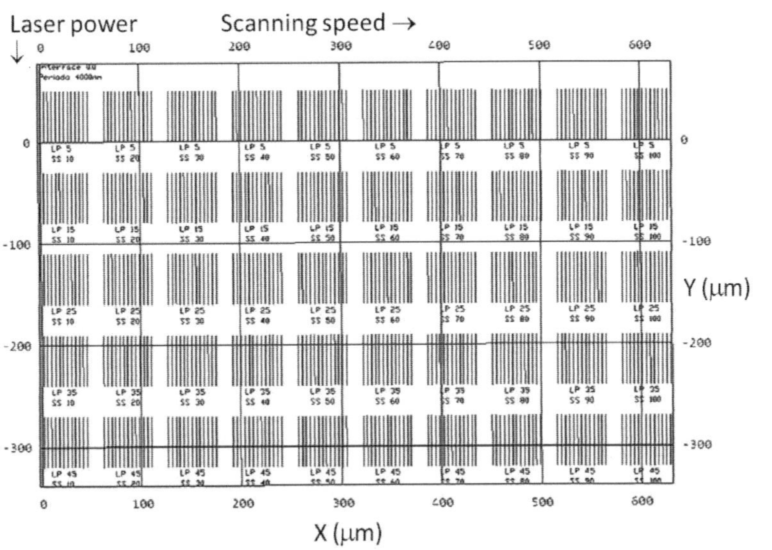

Figure 1. A scheme of the spatial organization of the gold arrays on the glass substrate. The laser power increases from top to bottom with a 5-mW step. The scanning speed increases from left to right with a step of 10 μm/s.

Raman spectra were recorded on a LabRAM micro-Raman spectrometer with a He-Ne laser excitation at 632.8 nm. The laser beam was focused onto the glass surface through microscope equipped with ×50 objective of a long working distance onto a spot of typical diameter of 5 μm, as shown in figure 2. Reference Raman spectra were taken from the substrate and the gold strips prior to any SERS measurements. After that Raman spectra were recorded with the sample immersed in a 10^{-5} M water solution of rhodamine-6G. Subsequently, the sample was washed in ethanol and exposed to a technical grade xylene (mixture of ortho-, meta-, and para- isomers) for 5 minutes, then washed again in ethanol, and finally Raman spectra were investigated for detection of trace amounts of xylene.

3. Experimental results and discussion

The reference spectrum, taken from the bulk of concentrated 10^{-3} M R6G solution, is shown in figure 3 (a). A typical spectrum taken from a gold strip, when the glass substrate was immersed in a 10^{-5} M R6G solution, is shown in figure 3 (b). The Raman lines of R6G significantly widen in the vicinity of gold strips and increase in intensity. When the integrated intensity of the Raman bands is normalized to the R6G concentration, as described in [1], the SERS enhancement factor (EF) accounts for ≈ 10^4

on average for the array written @22.5 mW/10 μm/s during the 2-PDLW process (the left-bottom array in figure 1).

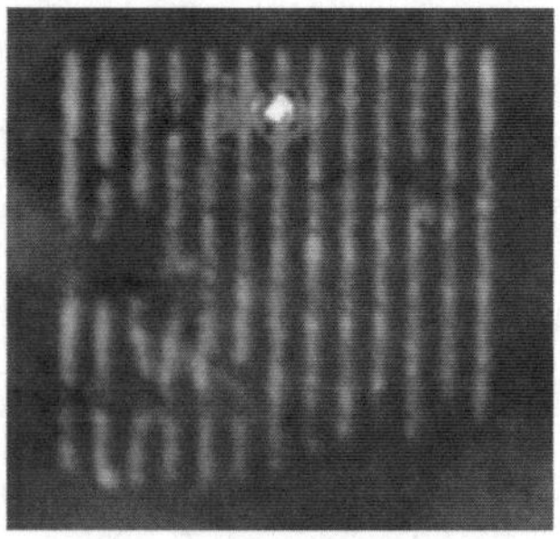

Figure 2. Gold strips as seen under optical microscope. The He-Ne laser focus on the sample is also visible as a bright spot.

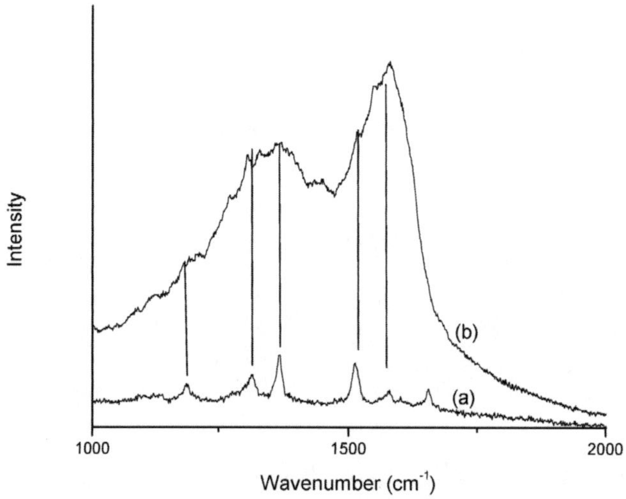

Figure 3. (a) A reference spectrum of a bulk 10^{-3} M R6G water solution. (b) Raman spectrum taken from a gold strip immersed in a 10^{-5} M R6G water solution. The intensity scale is the same for the two spectra.

Although a significant EF was evidenced for R6G, the quality of the SERS spectra is far from satisfactory due to the poor resolution of the spectral features. This fact could be explained by the strong enhancement of the wide photoluminescence (PL) background, which masks to a large extent the Raman signal. The physical origin of the PL enhancement is essentially the same as that of SERS, a SPR amplification of the local electric field around the metals surface [14]. According to [14] the PL background hampers the SERS signal at distances from the surface exceeding $\approx$ 10 nm. Thus, we may draw the conclusion that the R6G water solution does not penetrate effectively into the gaps between the gold nanoparticles, probably due to the large surface tension of water. Therefore, it would be instructive to test the SERS properties of our samples against organic solvents, which typically have about three times smaller surface tension than water.

We have chosen as a test substance the xylene, $C_6H_5(CH_3)_2$, which is a common organic solvent in many industrial applications. The toxicological hazard of xylene is well documented, which imposes specific regulations on its content in the common-life products [15]. Therefore, it could be of a definite practical interest to develop analytical methods for detection of trace amounts of xylene.

Figure 4 (a) shows a reference Raman spectrum of xylene, while spectra (b)-(e) were recorded on gold strips after exposure of the sample in xylene, as described in section 2. The Raman spectrum of bulk xylene overlays a wide PL background, which makes the Raman lines poorly resolvable. It is noteworthy that the spectrum (a) is multiplied by a factor of 5000 in order that its characteristic Raman features fit the intensity scale of the spectra (b)-(e).

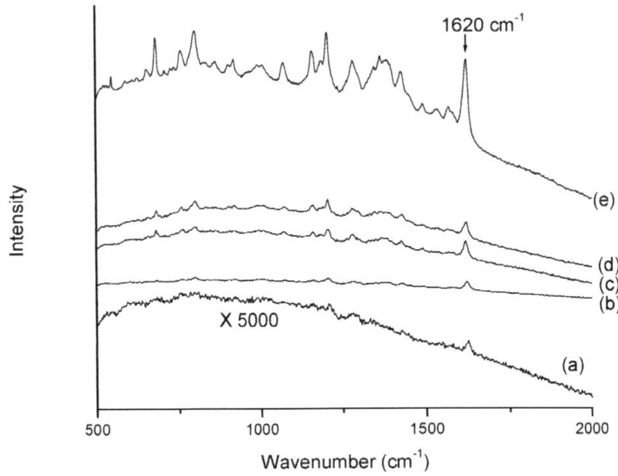

Figure 4. (a) Reference spectrum from bulk xylene; (b)-(e) SERS spectra recorded on gold strips written at a laser power of 22.5 mW and a scanning speed of 100, 70, 40, 10 μm/s respectively.

Therefore, the spectra presented in figure 4 evidence for a large SERS enhancement of xylene Raman features on the gold strips. The PL backgrounds enhances as well, but to a much lesser extent than the Raman spectrum. According to [14] this fact implies close proximity between xylene molecules and the gold nanoparticles.

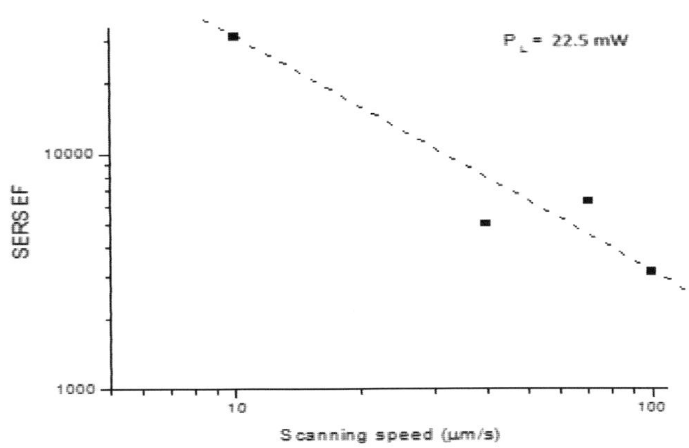

Figure 5. A log-log plot of the estimated SERS EF for the xylene line at 1620 cm^{-1} versus scanning speed, at a constant laser writing power $P_L = 22.5$ mW.

The SERS EF estimated for the strongest xylene line at 1620 cm^{-1} (see figure 4) correlates with the scanning speed (SS) in the process of laser writing, as shown in figure 5. The same order of magnitude EF have been estimated also for the other Raman lines present in the spectrum. Roughly, the SERS EF scales as 1/SS, i.e. in a proportion to the deposited laser energy per unit length. As reported in [13] the surface density of GNPs increases at smaller scanning speeds. Thus, our data correlate the SERS EF to the GNP density, which corroborates the assumption for formation of gap hot spots upon closer GNP packing.

4. Summary and conclusions

We established that the arrays of GNPs obtained by 2-PDLW process can serve as effective SERS substrates. We recorded SERS enhancement factors close or superior to 10^4 both for R6G solution and for xylene. The photoluminescence undergoes enhancement as well, which is comparable to the SERS enhancement in the case of R6G water solution. For xylene, however, the SERS signal overwhelms by two orders of magnitude the photoluminescence background. We attribute the different SERS behavior of the two analytes to the different surface tension of water and xylene. More concretely, the smaller surface tension of xylene allows the xylene molecules to fill-in the gap hot spots formed between closely packed GNPs.

We quantified that the SERS EF for xylene is maximized on GNP arrays written at highest laser power and smallest scanning speed. We explain this fact by the maximization of GNP density, and the formation of a large number of gap hot spots thereof, at highest laser exposures.

Acknowledgments

This work was sponsored by the Project INERA (GA316309) of the 7-th Framework Program of the European Union. Authors gratefully acknowledge also the scientific and technological support of the CNR user facility "Beyond Nano" – Pole of Cosenza, as well as the financial aid by the Bulgarian Scientific Fund through the grant FNI-TO2/26.

References

[1] Le Ru E C and Etchegoin P G 2013 *MRS Bull.* **38** 631
[2] Schatz G C, Young M A and Van Duyne R P 2006 *Top. Appl. Phys.* **103** 19
[3] More references Etchegoin P G and Le Ru E C 2011 *Surface Enhanced Raman Spectroscopy: Analytical, Biophysical and Life Science Applications* (Weinheim: Wiley-VCH) pp. 1–37
[4] Pettinger B, Picardi G, Schuster R and Ertl G. 2002 *Single Mol.* **5** 285–294
[5] Steidtner J and Pettinger B 2008 *Phys. Rev. Lett.* **100**, 236101–236104
[6] Prodan E, Radloff C, Halas N J and Nordlander P 2003 *Science* **302**, 419–422
[7] Etchegoin P, Cohen L F, Hartigan H *et. al.* 2003 *J. Chem. Phys.* **119**, 5281–5289
[8] Nafy L A 2012 *J. Raman Spectrosc.* **43** 1845
[9] Engelhardt S 2013 *Direct Laser Writing (Laser Technology in Biomimetics)* ed V. Schmidt and M. R. Belegratis (Berlin/Heidelberg: Springer-Verlag) pp. 13–56
[10] Bueckmann T *et. al.* 2012 *Adv.Mater.* **24** 2710–2714
[11] Yuan L, Ng M L and Herman P R 2015 *Opt. Mater. Express* **5** 515–529
[12] Ergin T, Stenger N, Brenner P, Pendry J and Wegener M 2010 *Science* **328** 337–339
[13] Ritacco T, Ricciardi L, La Deda M, and Giocondo M 2016 *J. Eur. Opt. Soc.*-Rapid **11** 16008
[14] Etchegoin P G and Le Ru E C 2010 *Basic Electromagnetic Theory of SERS (Surface Enhanced Raman Spectroscopy: Analytical, Biophysical and Life Science Applications)* ed S. Schlücker (Weinheim: Wiley-VCH Verlag GmbH & Co. KGaA) pp. 23–31
[15] Kandyala R, Raghavendra S P C and Rajasekharan S T 2010 *J Oral Maxillofac Pathol.* **14** 1–5